IMAGES
of America

THE DIXIE HIGHWAY IN ILLINOIS

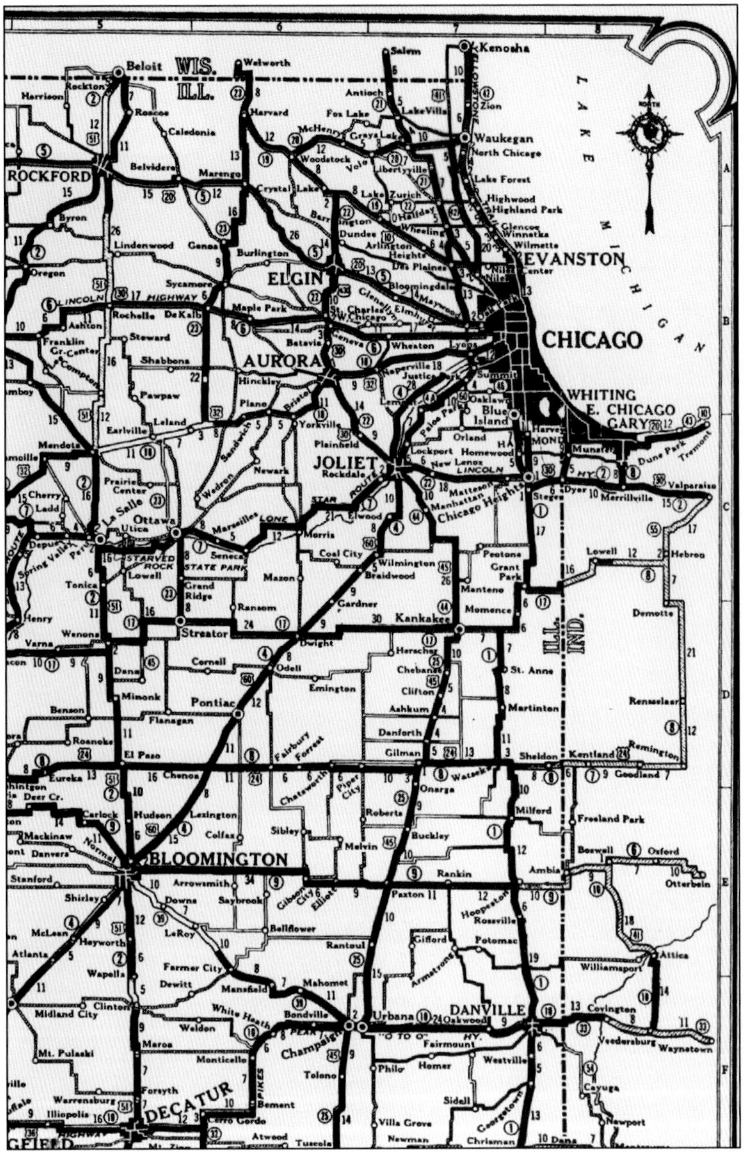

This Rand McNally map from 1926 gives a perfect overview of the Dixie Highway in Illinois. The route of the highway can be traced by following the heavy black line from Blue Island through Homewood to Chicago Heights. South of Chicago Heights, the highway is marked as Illinois Route 1 to Danville. The highway then followed Illinois Route 10 east into Indiana. By the mid-1920s, paved roadways were proliferating; as the map illustrates, however, the state's highway network was still in its infancy. (Courtesy Danville Public Library.)

On the cover: The southern point of the Dixie Highway in Illinois is the city of Danville. The highway followed Vermilion Street south to Main Street then east to the state line. This view of Main Street, looking west toward Vermilion Street, shows downtown Danville and the Dixie Highway bustling with activity. The First National Bank building, the tallest in the city, is shown at right. The view also shows an early traffic signal located in the center of the interurban tracks. (Courtesy Vermilion County Museum.)

IMAGES
of America

THE DIXIE HIGHWAY IN ILLINOIS

James R. Wright

ARCADIA
PUBLISHING

ISBN 978-0-7385-6002-1

Published by Arcadia Publishing
Charleston, South Carolina

Printed in the United States of America

Library of Congress Control Number: 2009921934

For all general information contact Arcadia Publishing at:
Telephone 843-853-2070
Fax 843-853-0044
E-mail sales@arcadiapublishing.com
For customer service and orders:
Toll-Free 1-888-313-2665

Visit us on the Internet at www.arcadiapublishing.com

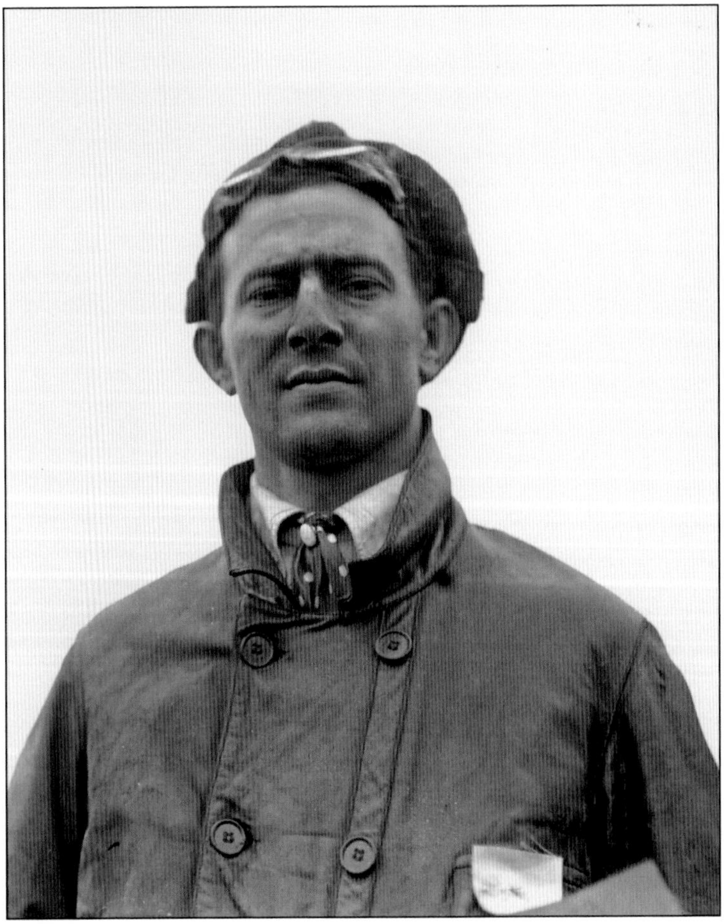

Carl G. Fisher, an Indiana native, has been dubbed the "father of the Dixie Highway." He felt the Dixie Highway would be a perfect north–south route bringing northerners to the sunny climes of Florida. He helped organize the Dixie Highway Association in 1915 and was an organizer of the Lincoln Highway Association in 1913. Fisher was also involved in developing the Indianapolis Motor Speedway and raced cars himself. (Courtesy Chicago History Museum.)

Contents

Acknowledgments

Early credit and thanks go to Homewood village president Richard Hofeld, who had a vision to promote the history of the Dixie Highway, and to chairperson Elaine Egdorf and the members of the Homewood Heritage Committee who refined this vision and enlisted representatives from communities all along the highway in Illinois. From these beginnings in 1999, Drivin' the Dixie, an event that takes place each June, was born. Many thanks go to others involved in Drivin' the Dixie for their efforts promoting the highway, including the A's R Us Model A Ford Club, particularly Bob Steinmetz and Wally Ekstrom, and to the Dixie Highway community organizers for their work in ensuring this event is as successful as it is.

Many of these individuals have served as a foundation for support, information, and photographs for this book. Others who have lent their assistance are numerous, and I will try to recognize many of them here: Rob Medina, Chicago Historical Society; Carol Flynn and Linda Lamberti, Ridge Historical Society; Richard Reuss; Mike Janko, Blue Island Historical Society; Jerry Bosch and George Peso, Evergreen Park Historical Commission; village president Donald Schupek and trustee Charles Spindler, Village of Posen; Dave Bartlett; Geraldine Leuther Justice; Homewood Historical Society; Chuck Hecht; Carolyn Wagner, Chicago Heights Public Library; Dominic Candeloro; Mike Leoni; Richard Felicetti and Don Croarkin, Bloom Township High School Archives; James Leech, South Chicago Heights Historical Committee; Millie Zeman, Steger Historical Society; Miriam and William Giebel; Carol and Wehlan Triebold; Phyllis Monks, Crete Historical Society; Virginia Bath and Arnold Cooper, Beecher Historical Society; Marilyn Ohlendorf; Sylvia Unruh; Janet Nelson; Victor Frahm, Ileane Breslin, and Vern Koelling, Grant Park Historical Society; Harold and Amelia Schriefer; Landis Wehrmann; Kevin McNulty; Janine Loftus, Main Street Momence; Jean Stetson, Edward Chipman Public Library–Momence; Brent Langman; Richard McHie; Robert Hoekstra, Don Bonvallet; Gilbert Irps; Robert Bovie; Doug Redenius; Fr. James Fanale, St. Anne Church; Marilyn Miller; Jeff Webster; Ginny Lee and Judy Ficke, Iroquois County Historical Society; Deborah LaFine, Iroquois County Genealogical Society; Jim Pence; John Devine; Gloria Lynch; Sue Crist, Milford Public Library; Frank and Fred Hines; Jean Minnick and Carol Hicks, Hoopeston Historical Society; Lou Graham, Hoopeston Public Library; Kevin Young, Rossville Historical Society; John Brain, Newell Township Historical Society; Norma Clem Bailey; Sue Richter, Vermilion County Museum; Roberta Allen and Jackie Woodard-Smith, Danville Public Library; secretary Milton Sees, Dennis Markwell, Joseph Putnam, Amy Eller, and Eugene Smania, Illinois Department of Transportation, and Ed Smith. For those whom I may have missed, please accept my sincerest apologies.

Finally, I would like to thank Karen Smith for editorial assistance and my good friend Phil Serviss for his encouragement, technical advice, and research assistance.

Introduction

The origins of the Dixie Highway in Illinois date to the 1820s, when Gurdon Saltonstall Hubbard, a young clerk with John Jacob Astor's American Fur Trading Company, arrived in eastern Illinois. Hubbard set up trading posts between Chicago and Vincennes, Indiana, and the route between these became known as Hubbard's Trail. In 1835, the Illinois General Assembly ordered that a road be established and marked between Vincennes and Chicago. Hubbard's Trail was selected as the most direct route. This was the first official state route established in Illinois and was known as the Chicago-Vincennes Road, later known as Illinois Route 1. Some 70 years later, the path of Hubbard's Trail—the Chicago-Vincennes Road—would serve as the foundation for the Dixie Highway.

To appreciate the importance of the Dixie Highway in Illinois, it is necessary to turn the hands of time back over 100 years. At the beginning of the 20th century, travel was by horse and carriage, and trips were confined generally to a few miles. Longer trips were made by train, and people had to adhere to railroad timetables and routes to make these trips possible. With the new century, things changed rapidly. Technological developments in transportation, particularly the automobile, allowed more freedom in planning and making both short and longer distance trips. An impediment to this increased freedom, however, was the lack of improved roadways. Most roads were primarily dirt roads, and while passable in fair weather, they quickly became rutted mud holes after a few hours of rain.

To improve conditions, organizations such as the American Automobile Association lobbied for new road construction. Farmers' organizations too became advocates of road construction to help lower the costs of marketing their farm products. The establishment of rural free delivery became a major influence in the federal government also promoting the construction of higher quality stone roads to replace the dirt paths its carriers encountered on rural routes.

These groups and others coalesced into the Good Roads Movement, which lobbied local, state, and federal legislatures for improved roadways throughout the country. Improved roads generally ended at city or town limits, and the Good Roads Movement wanted to extend the brick, macadam, or concrete roads past these borders to benefit commerce, agriculture, and tourism. The development of the Dixie Highway as the major route linking the South with the midwestern North was a product of this movement.

One individual strongly associated with the development of the Dixie Highway, and dubbed the "father of the Dixie Highway," was Carl G. Fisher. An Indiana native and entrepreneur, he owned the Prest-O-Light Company, which manufactured compressed acetylene gas headlight systems that made night driving possible. Fisher was an organizer of the Indianapolis Motor Speedway and had extensive real estate holdings in what was to become Miami Beach. This combination of interests in selling more cars and promoting and selling real estate in Florida led Fisher to devote considerable energy to the development of better roads and to become a major force in promoting the automobile for tourism and specifically for north–south tourism. Fisher had been involved in the Lincoln Highway project in 1913, where he developed connections helpful for the Dixie Highway project.

In November 1914, Fisher and W. S. Gilbreath, secretary of the Hoosier Motor Club, attended the American Road Congress in Atlanta. In attendance were the governors of Georgia and Indiana. Fisher and Gilbreath addressed the conference and told of their vision for a north–south highway, which they said would generate an immense amount of money that would be left in the South "by a peaceful invasion by desirable people from the north." With that bit of salesmanship, their idea took off, and the Dixie Highway was born.

For the next several months Fisher traveled to Georgia, Tennessee, and Kentucky promoting the idea of the highway, and in April 1915, the Dixie Highway Association was formed in Chattanooga, Tennessee. At the organizational meeting, over 5,000 people from over 100 communities in the states of Florida, Georgia, Tennessee, Kentucky, Ohio, Indiana, and Illinois attended. Competition between communities for a place on the highway was fierce.

Extensive debate about the location of the highway and pressure from the State of Michigan to be included on the route led to a decision in May 1915 to designate two routes of the highway. A western division carried travelers from Chicago into the South via Indianapolis, Nashville, and Atlanta, while an eastern division had its northern terminus in Sault Ste. Marie, Michigan, and took travelers through Detroit, Toledo, Cincinnati, and Lexington on their way to the highway's southern terminus in Miami Beach. Later a branch of the highway extended into North Carolina and South Carolina, and the northern terminus of the western division also became Sault Ste. Marie.

Between 1915 and 1927 (when the Dixie Highway Association disbanded), nearly 5,786 miles of roadway had been improved on the two divisions of the highway. This was double the length that organizers originally anticipated. Gravel and poor-quality asphalt roads were changed to brick or concrete. Improvements were paid for by the local community or by county or state funding. In some areas, contributions from business owners along the route, such as hotels, restaurants, and service stations, also helped fund the roadway improvements.

The construction of the Dixie Highway nationally was complete by 1926. About this time, the federal government established a highway numbering system. Highways were now numbered rather than named and were marked by standardized black-and-white signs. The different shaped and colored signs of the named highways such as the white, red, and white bands with a DH in the center of the signs marking the Dixie Highway disappeared. The Dixie Highway became known as U.S. Routes 41, 25, and 27 and by other U.S. and state route numbers depending on which state one was in. In most of Illinois, the road became known as Illinois Route 1.

The selection of the route in Illinois, like other states, was subject to debate and some contention. Fisher wanted the highway to start in Chicago, but, as a Hoosier, he envisioned the highway leaving Chicago's southeast side and quickly entering Indiana. Under Fisher's plan, Chicago would be the only Illinois city on the highway. Illinois delegates to the 1915 route selection meeting would not have this. They lobbied delegates from other states to support a greater role for "the Land of Lincoln," and their efforts paid off.

The route finally selected called for the Dixie Highway in Illinois to run about 136 miles from Chicago to Danville. The route started at Adams Street and Michigan Boulevard in Chicago and followed Michigan south to 55th Street, west to Western Avenue, and south on Western through Blue Island to the present route of the Dixie Highway through Chicago Heights, following present-day Illinois Route 1 through Crete, Beecher, Grant Park, and Momence.

South of Momence, at Kinney's Corner, the highway continued due south and followed existing roads that paralleled the Chicago and Eastern Illinois Railroad tracks through the hamlets of Koster and Wichert. The highway then proceeded south on what is now 8500E Road then west on 6750S Road, joining Station Street through downtown St. Anne. The road then followed the highway's current alignment to Guertin Street and continued south on 7500E and Cable Line Roads into Papineau, then west to where Illinois Route 1 is located today.

In Iroquois County, today's alignment of the highway varies from the original route south of Martinton, at Pittwood, and south of Milford to eliminate 90-degree turns that were so dangerous to early motorists traveling at higher speeds. Similar refinements were made at the Iroquois-Vermilion County line north of Hoopeston. Many of these sharp turns were notorious for fatal accidents, earning the intersections chilling names like "Dead Man's Curve."

A dedication motorcade traveled the route from Chicago to Miami in October 1915 as a means of promoting the highway. A wide variety of road conditions were encountered, but some of the best were in Illinois. Communities all along the route in Illinois cheered motorcade participants as they passed through their towns. The group traveled from the Art Institute of Chicago to Danville on the first day and was feted at the Elks Club with a sumptuous banquet that night.

Construction on improving the road in Illinois started in 1916. Over the next five years, paving work progressed on the highway until the last section to be paved between Crete and Beecher was completed in May 1921. For much of the route, the finished surface was concrete pavement with the exception of stretches in Vermilion County that were paved with brick. The Dixie Highway, when completed, was the longest continuous paved road in Illinois at the time.

To mark the completion of the highway in Illinois, another motorcade occurred on August 12, 1921, starting in Danville and ending in Chicago at the Pageant of Progress Exposition. Chicago mayor William "Big Bill" Thompson declared the day Dixie Highway Day, and ceremonies took place at Municipal, now Navy, Pier. Each town the motorcade passed through followed with the girl chosen to represent their community and a contingent of citizens in their automobiles. Another less-publicized motorcade took place on the 10th anniversary of the dedication of the highway route in October 1925.

By the 1930s, the Dixie Highway in Illinois needed modernization. State and county highway departments worked on plans to widen narrow stretches and eliminate sharp turns with safer, sweeping curves. Much of the interstate traffic that had once traveled the Illinois highway was now traversing routes like U.S. Route 41 in Indiana, a modern highway that bypassed the communities along its route allowing motorists to quickly travel to their destination without the delays necessitated by stop-and-go traffic in cities and towns.

In 1937, proposals were advanced by Robert Kingery and the Chicago Regional Planning Association to build a new stretch of the highway leading directly from the intersection of Illinois Route 17 and the Dixie Highway, east of Grant Park, southward through Iroquois County, passing west of Beaverville, and then directly south through Milford to Danville. As proposed, this route would save 17 miles of travel and provide a more direct route through the eastern part of the state. The route would have eliminated Grant Park, Momence, St. Anne, Martinton, and Watseka from the Dixie Highway.

As expected, there was a great deal of opposition to this proposal in Kankakee and Iroquois Counties. This opposition, along with the intervening war years, shelved the plan for about a decade. It was resurrected again in the late 1940s but never gained much traction. With the advent of the Interstate Highway System and the completion of Interstate 65 in Indiana and Interstate 57 in Illinois, the highway would never again regain its stature as a preeminent link from north to south in Illinois.

Today, although few travelers bound for Florida motor down the Dixie Highway in Illinois, the road remains the backbone of the communities it passes through. This book is meant to give a glimpse into the history of this highway and the towns along it.

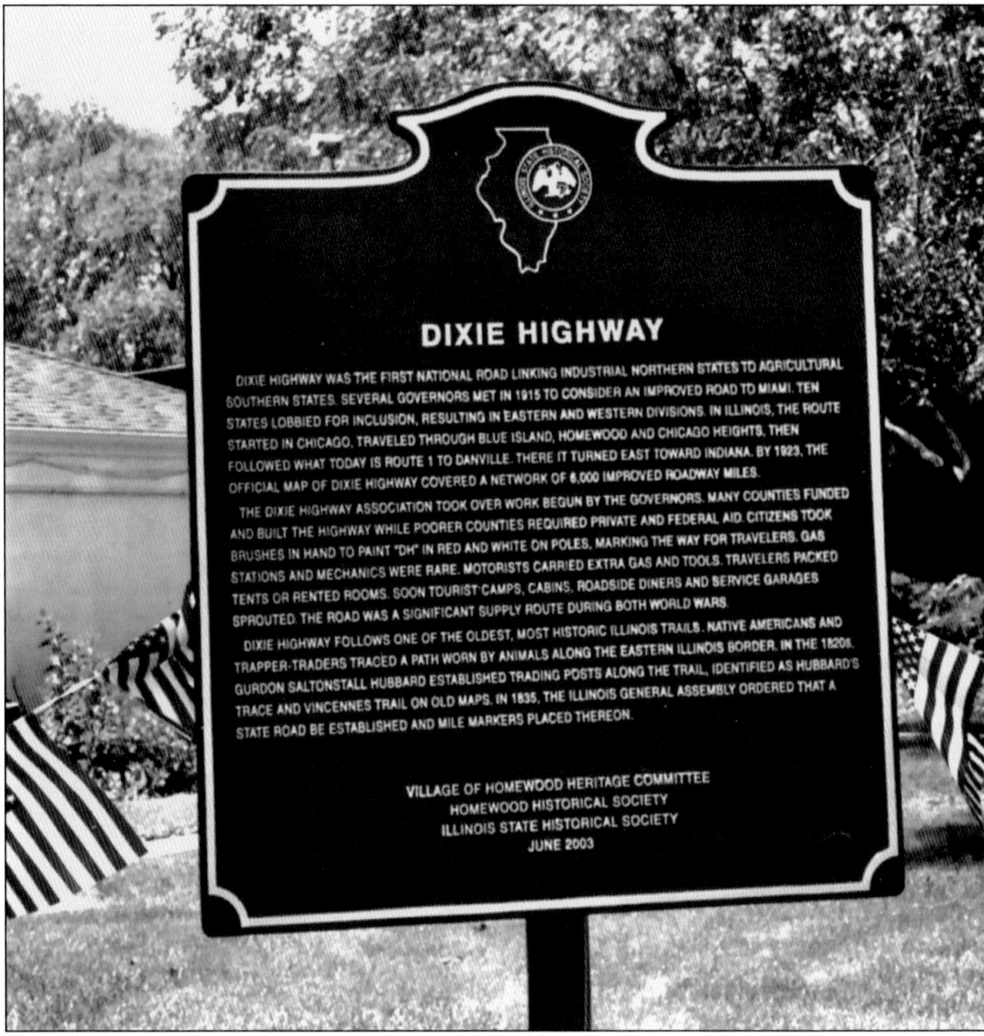

This marker sponsored by the Homewood Historical Society, the Village of Homewood Heritage Committee, and the Illinois State Historical Society was dedicated on June 1, 2003, to commemorate the history and importance of the Dixie Highway in Illinois. The marker is located in Independence Park at the corner of the Dixie Highway and Hickory Road in downtown Homewood. (Author's collection.)

One

Cook County

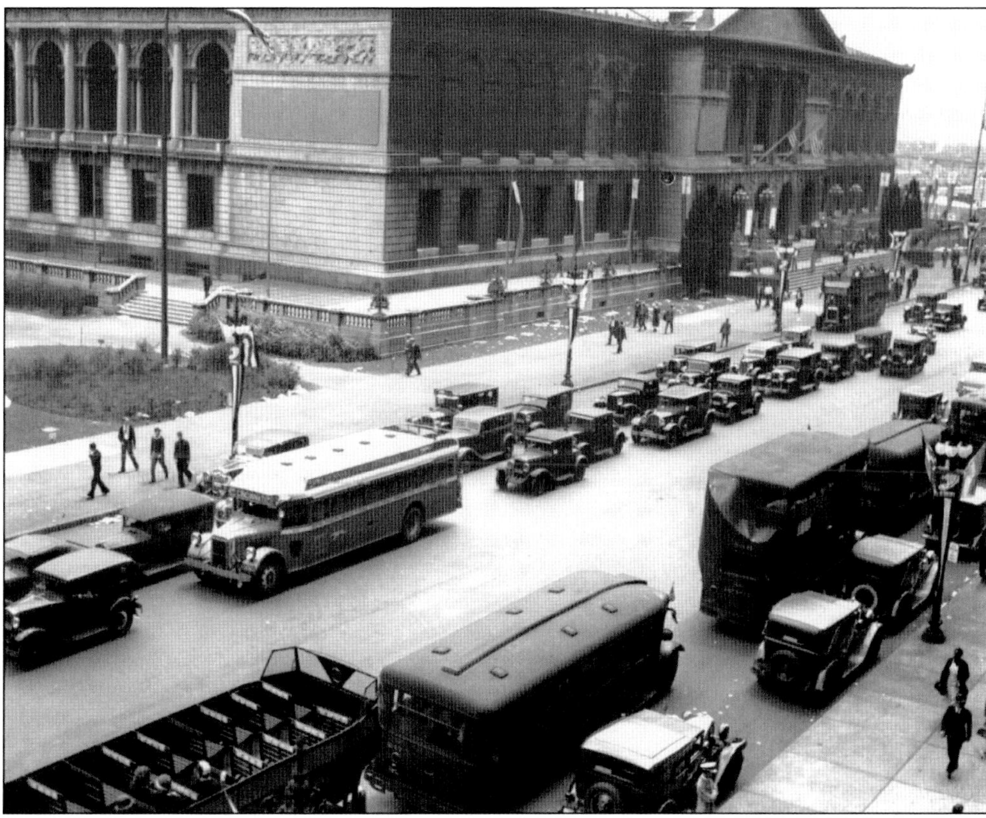

The Art Institute of Chicago on Michigan Avenue at Adams Street was considered to be the starting point of the Dixie Highway in Chicago. The institute played host to the ceremonies marking the start of the Dixie Highway inspection tour in 1915 and, as this photograph illustrates, has always been a popular attraction for tourists visiting the city. The museum was founded in 1879, and the current building dates to 1893. (Courtesy Chicago Public Library.)

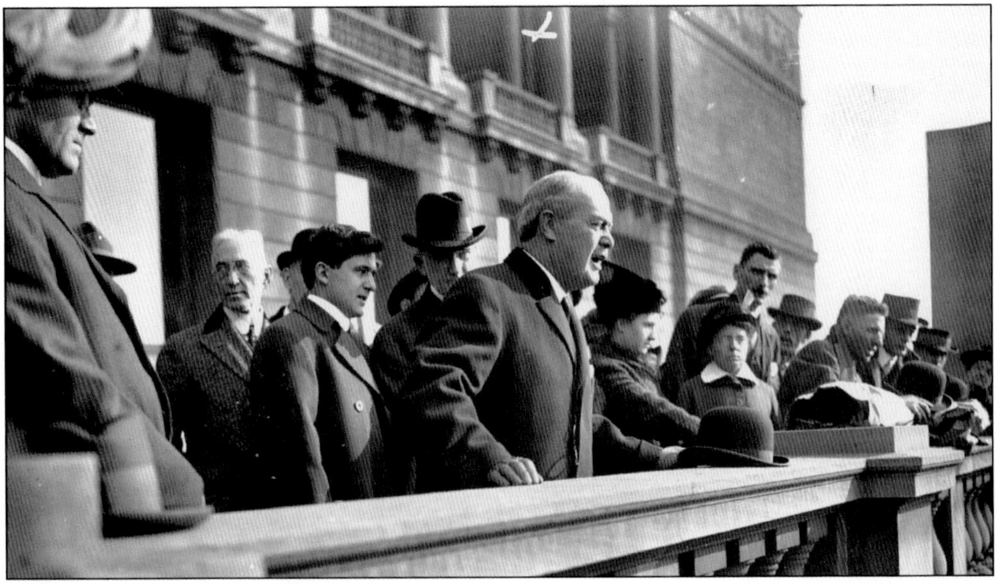

In October 1915, Carl G. Fisher and other Dixie Highway boosters embarked on an inspection trip of the highway's route traveling from Chicago to Miami. The trip started early on Saturday, October 9, with a program held at the base of the Spirit of the Lakes fountain on the grounds of the Art Institute of Chicago. Gov. Edward Dunne gave a rousing speech before joining the motorcade on its trip through Illinois. (Courtesy Chicago History Museum.)

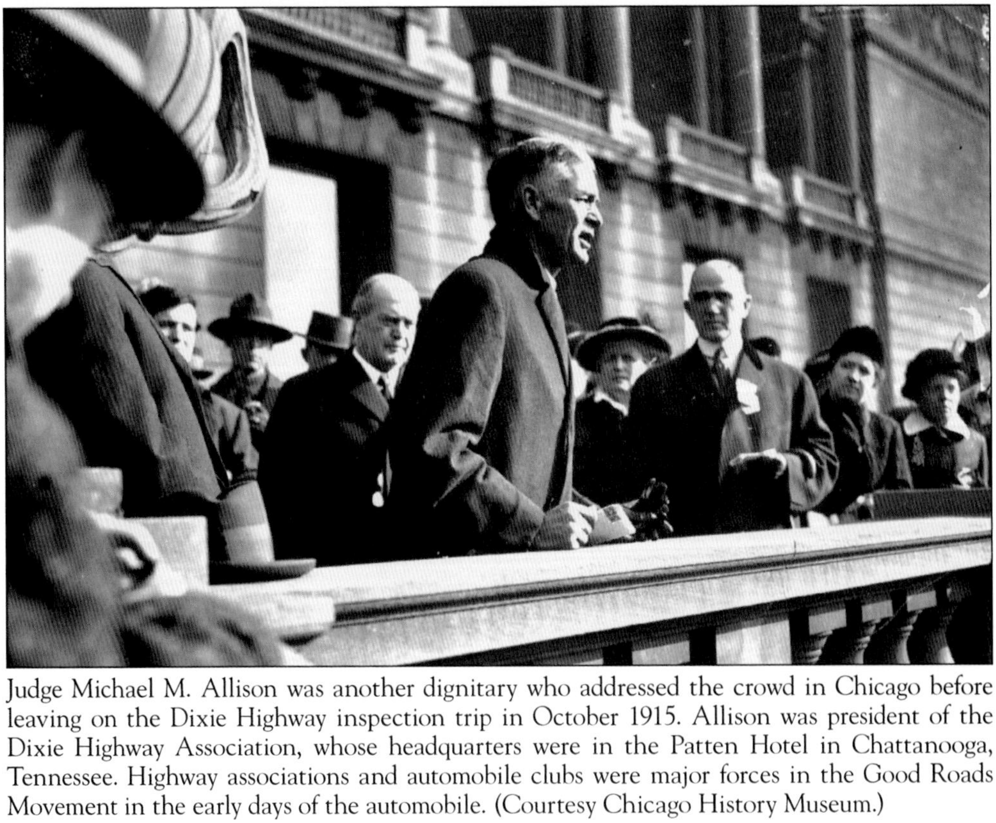

Judge Michael M. Allison was another dignitary who addressed the crowd in Chicago before leaving on the Dixie Highway inspection trip in October 1915. Allison was president of the Dixie Highway Association, whose headquarters were in the Patten Hotel in Chattanooga, Tennessee. Highway associations and automobile clubs were major forces in the Good Roads Movement in the early days of the automobile. (Courtesy Chicago History Museum.)

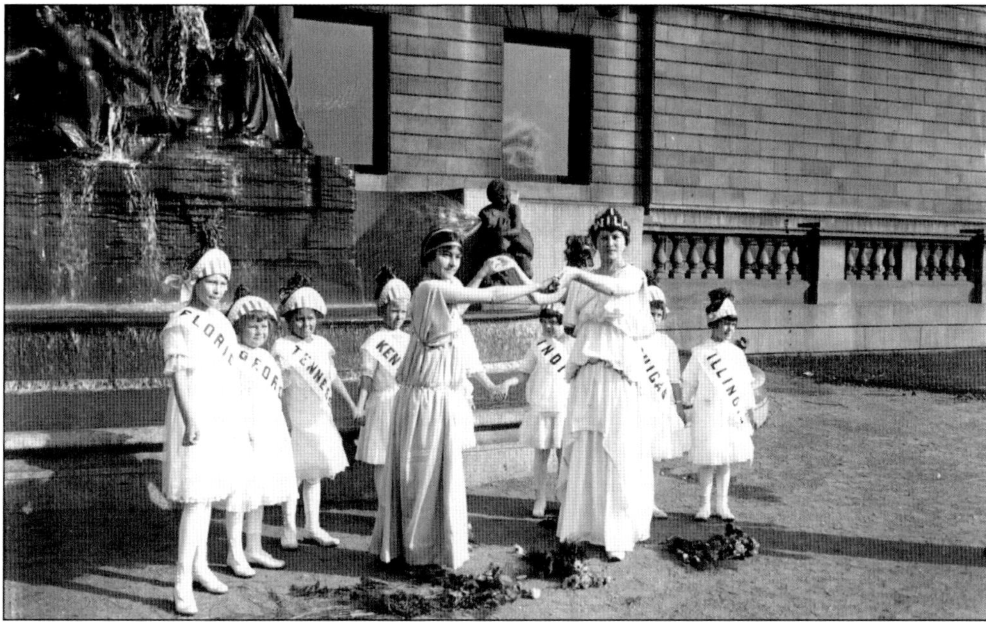

As part of the October 1915 dedicatory program at the Art Institute of Chicago, young ladies representing Miss Dixie and Miss Chicago join hands in a show ceremoniously linking the North and South. Young girls representing Florida, Georgia, Tennessee, Kentucky, Indiana, Michigan, and Illinois, the states the highway passes through, surround them in front of the Spirit of the Lakes fountain. (Courtesy Chicago History Museum.)

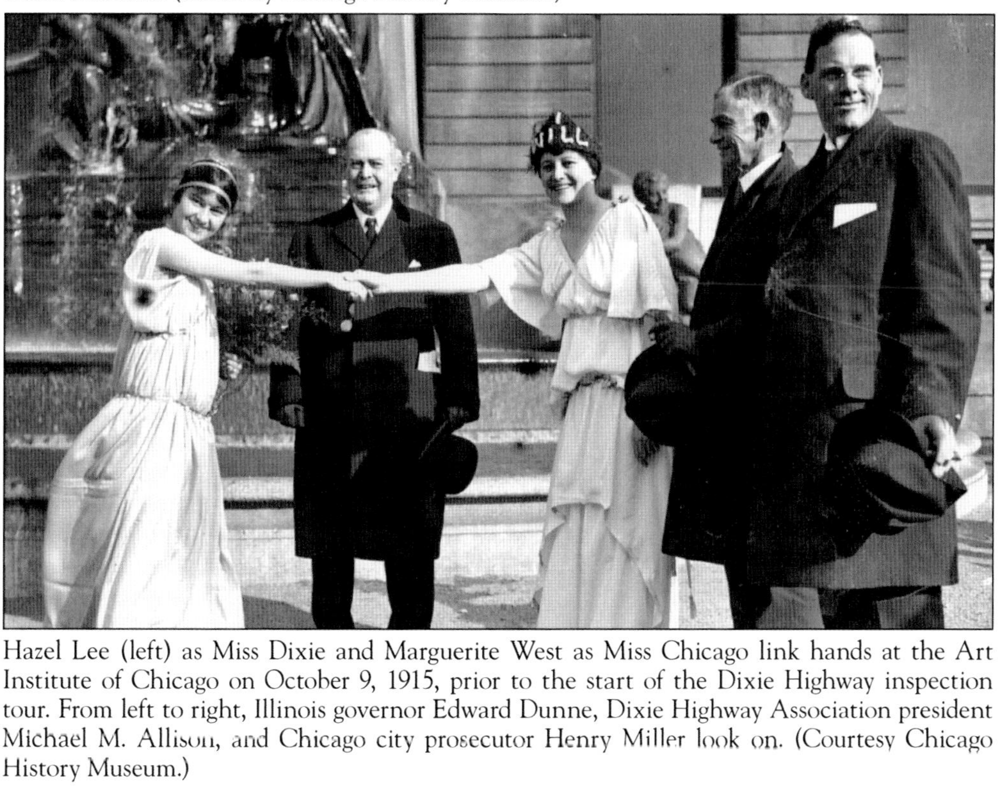

Hazel Lee (left) as Miss Dixie and Marguerite West as Miss Chicago link hands at the Art Institute of Chicago on October 9, 1915, prior to the start of the Dixie Highway inspection tour. From left to right, Illinois governor Edward Dunne, Dixie Highway Association president Michael M. Allison, and Chicago city prosecutor Henry Miller look on. (Courtesy Chicago History Museum.)

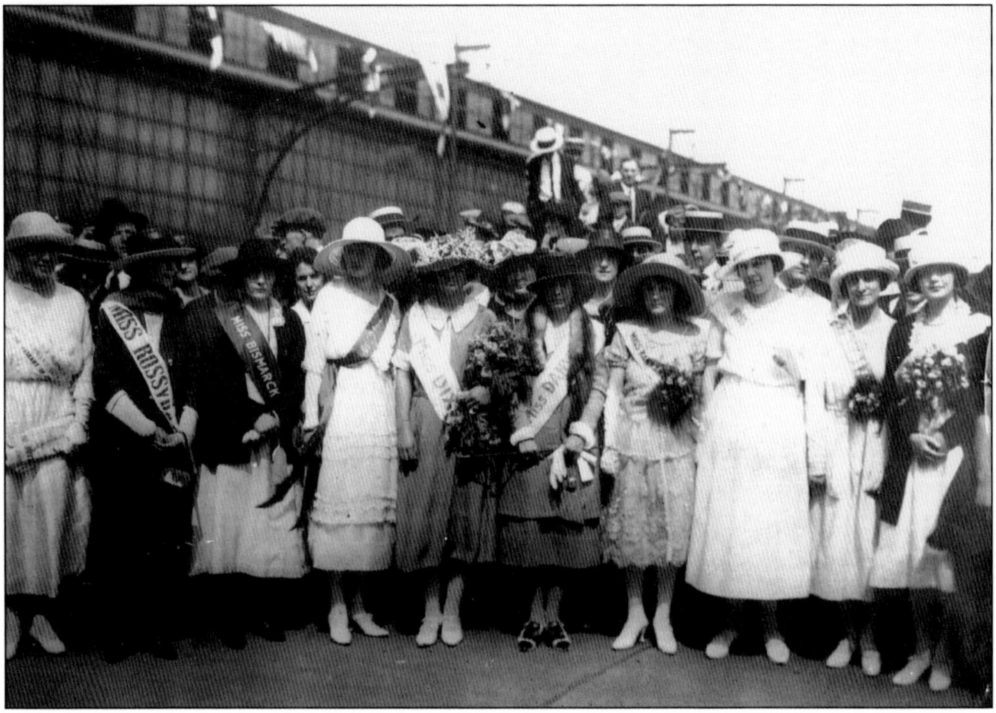

After paving on the Dixie Highway was completed, Illinois Dixie Highway communities designated August 12, 1921, as the "official" opening of the highway. A parade of automobiles left Danville for Chicago at 6:00 a.m. and, at each community on the way, other automobiles joined the caravan along with their local "Miss." Miss Dixie and other young ladies representing Dixie Highway communities pose for this photograph at Chicago's Municipal (now Navy) Pier. (Courtesy Chicago History Museum.)

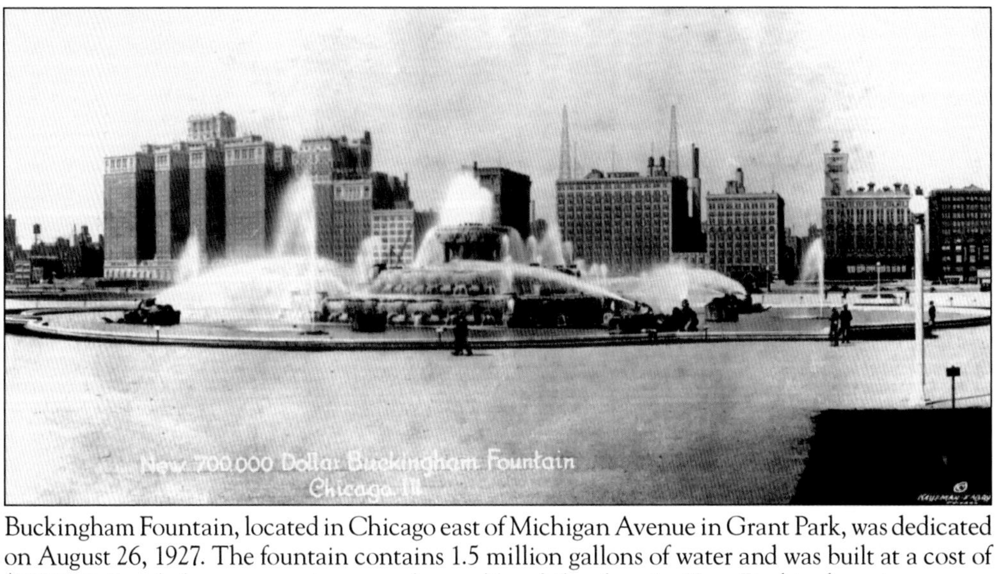

Buckingham Fountain, located in Chicago east of Michigan Avenue in Grant Park, was dedicated on August 26, 1927. The fountain contains 1.5 million gallons of water and was built at a cost of $700,000. Funds for construction were donated by philanthropist Kate Buckingham in memory of her late brother Clarence. The fountain became an instant attraction for tourists in Chicago on the Dixie Highway. (Author's collection.)

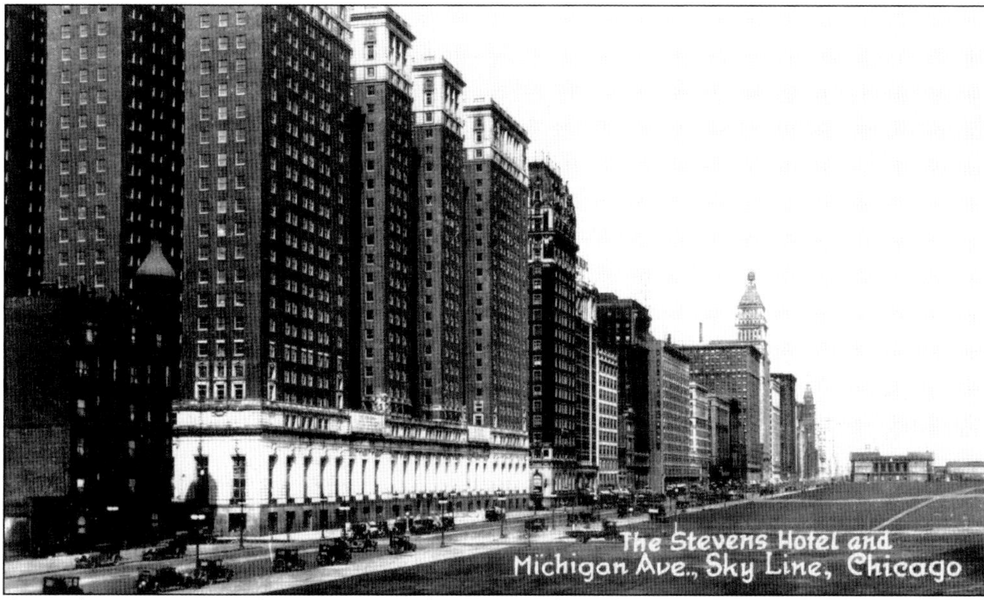

Opened on May 2, 1927, the Stevens Hotel at 720 South Michigan Avenue, on the Dixie Highway in Chicago, was the largest hotel in the world at the time. The 28-story structure occupied an entire city block and contained 3,000 guest rooms. The onset of the Depression spelled hard times for the hotel, which was used as military housing during World War II. After the war, Conrad Hilton bought the property and named it for himself; it remains one of the finest in the city. (Author's collection.)

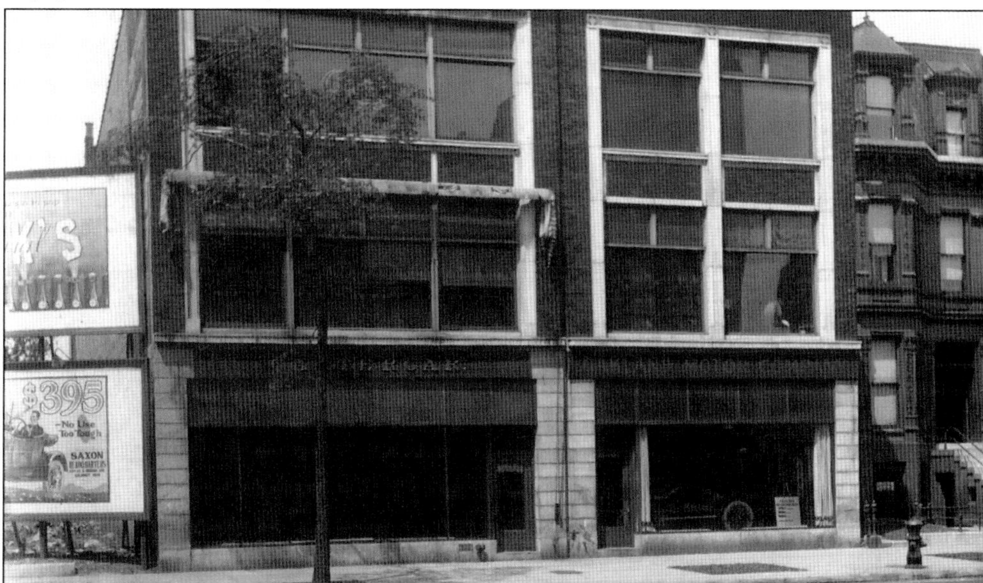

The Dixie Highway and Michigan Avenue south of Twelfth Street stretching to Twenty-sixth Street was known as Motor Row in Chicago. Motor rows were common in large cities where car and related companies sought to create districts where customers had easy access to automobile dealerships and repair firms. This concept is not unlike the automobile mall of today. The Oakland Motor Car dealer at 1840 South Michigan Avenue is pictured. (Courtesy Chicago History Museum.)

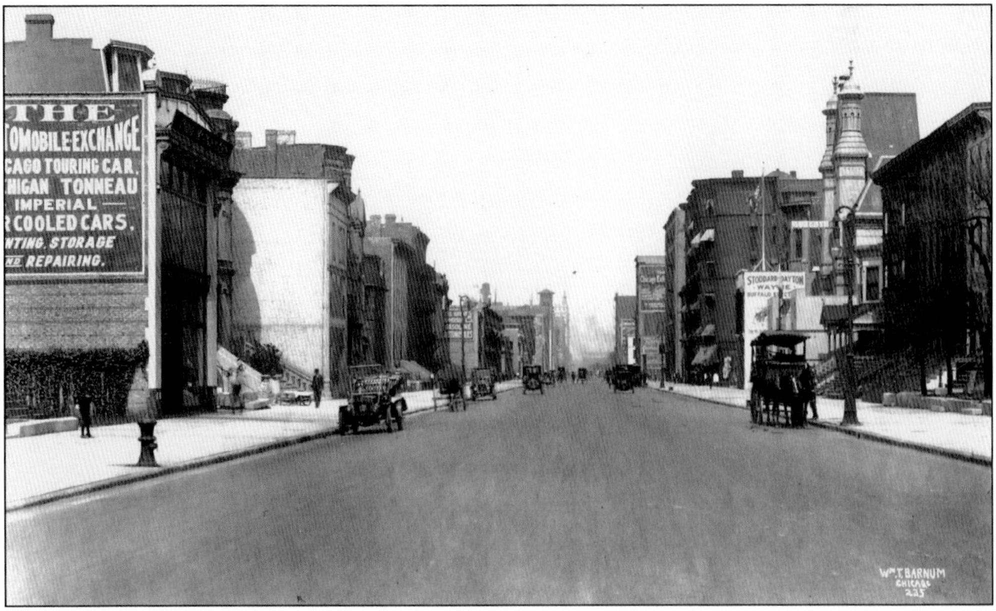

Traffic on Michigan Avenue and Motor Row in Chicago early in the second decade of the 20th century was a mix of horse carriages and horseless carriages. This view looking north from Eighteenth Street shows the rising popularity of the automobile evident by the traffic on the street and a number of advertising signs painted on many of the buildings. (Courtesy Chicago History Museum.)

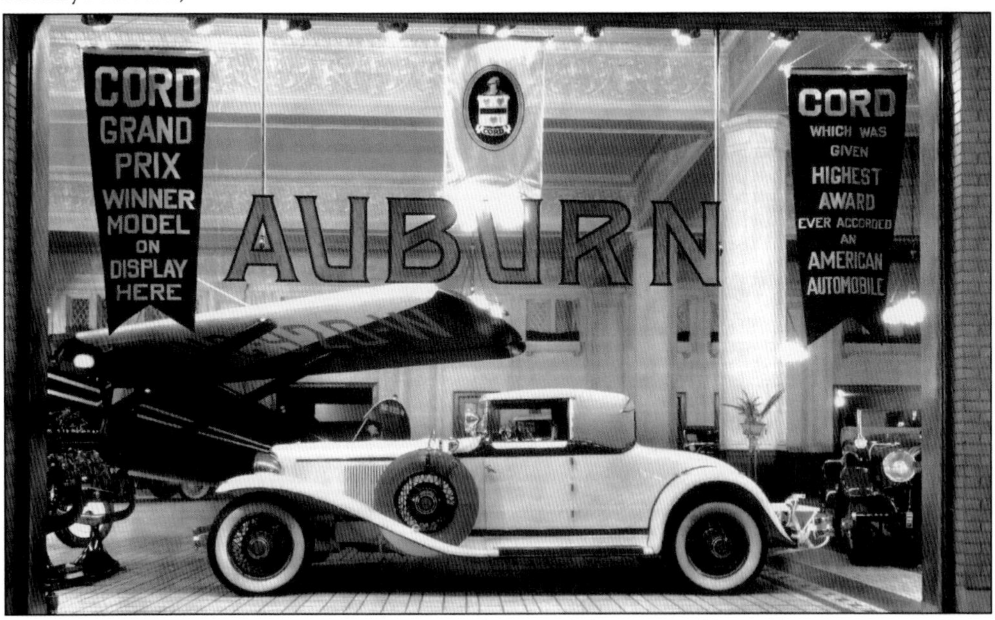

At its peak, as many as 116 different makes of automobiles were sold on Chicago's Motor Row. By the 1920s, dealer showrooms became larger and more opulent. The Auburn-Cord dealership at 2401 South Michigan Avenue displayed not only the company's automobile models but also a plane of the Stinson Aircraft Corporation, a subsidiary company. The Cord L-29 model centered in the display was the first American front-wheel drive car offered for sale to the public. (Courtesy Auburn-Cord-Duesenberg Museum.)

The Lexington Hotel, located at 2135 South Michigan Avenue, was another landmark hotel on the Dixie Highway route in Chicago. Completed in 1892 in time for the World's Columbian Exposition, the hotel gained a more lasting and unsavory reputation as the headquarters of Al Capone and his gang from 1928 to 1932. Eventually converted to a low-rent residential hotel, the building was demolished in 1995. (Courtesy Chicago History Museum.)

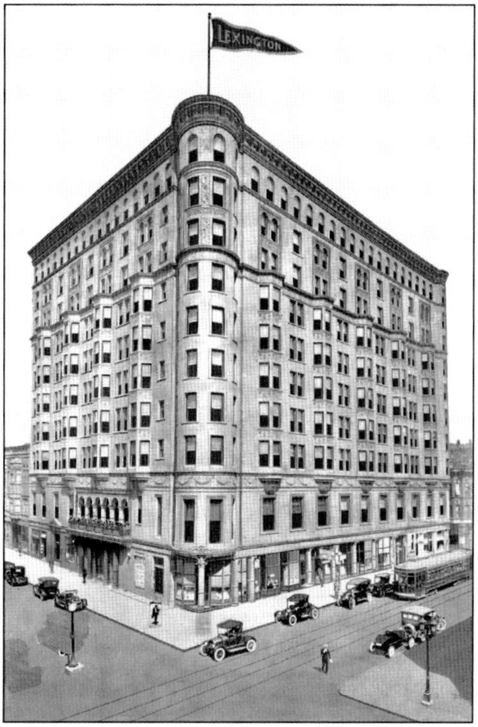

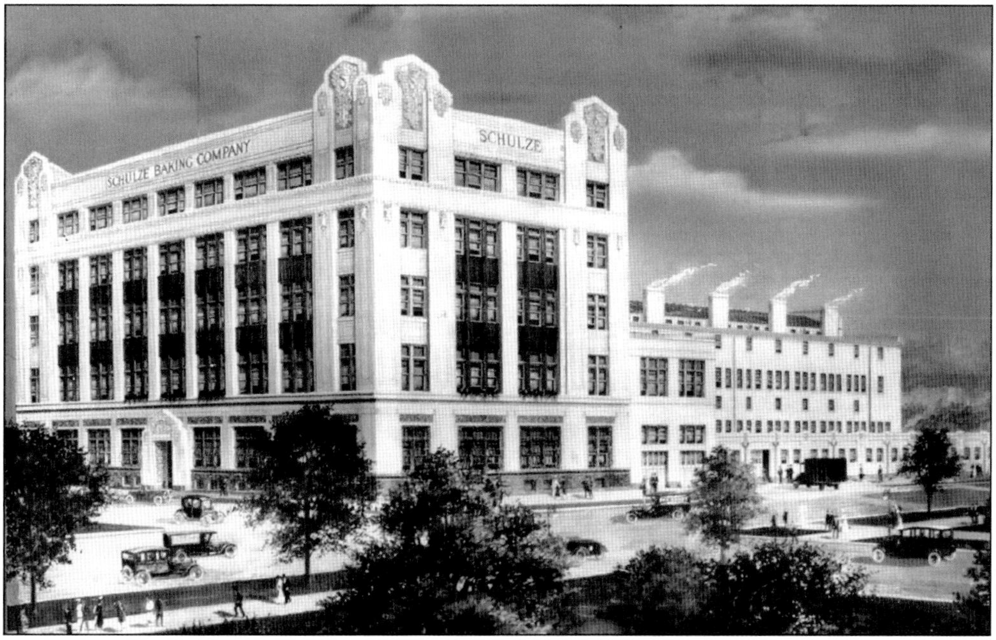

After traveling south on Michigan Avenue, the route of the Dixie Highway in Chicago proceeded west on Fifty-fifth Street, Garfield Boulevard, to Western Avenue. West of Michigan Avenue on Fifty-fifth Street, the Schulze Baking Company, makers of Butternut Bread, constructed this modern plant in 1914. Interstate Baking Company acquired the company in 1937. The landmark building still stands but has been closed since 2004. (Courtesy Interstate Bakeries Corporation.)

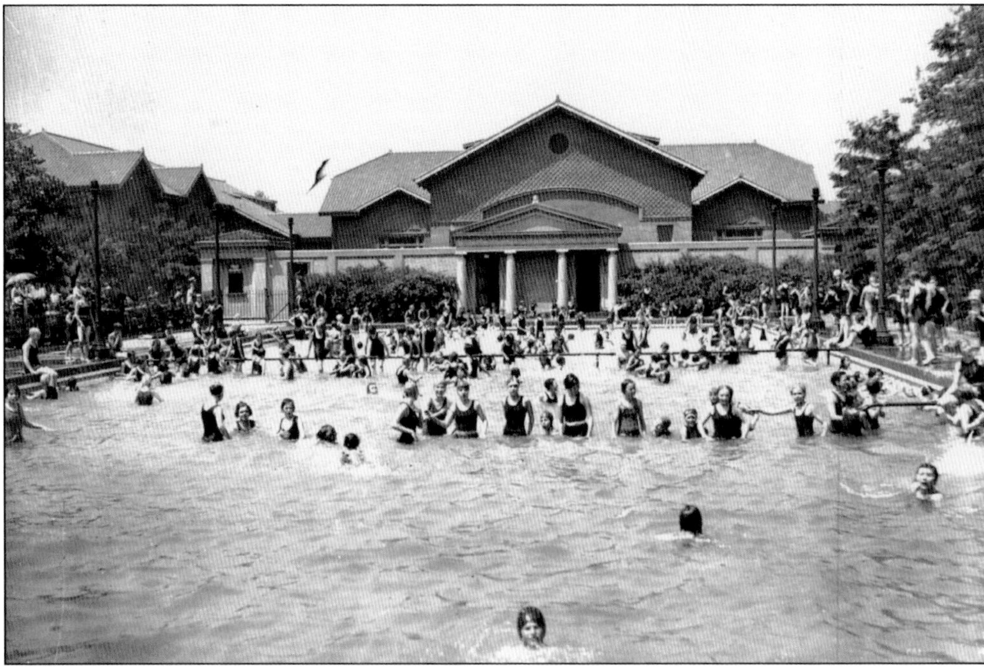

In Chicago, the Dixie Highway traveled west on Garfield Boulevard, Fifty-fifth Street, to Western Avenue, where the highway turned south and followed Western Avenue into Blue Island. Gage Park and its pool and field house, located at the intersection of Garfield Boulevard and Western Avenue were, and still are, landmarks for travelers in this area. The park's field house was built in 1926. (Courtesy Chicago History Museum.)

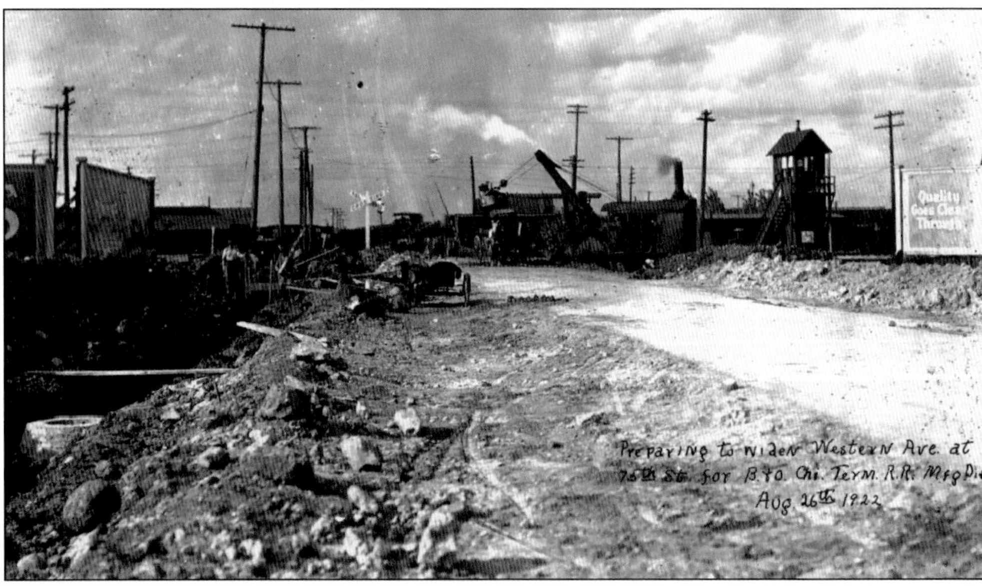

In 1922, a massive project was undertaken to widen Western Avenue, the Dixie Highway, in the area of the Baltimore and Ohio Railroad tracks near Seventy-fifth Street in Chicago. The project extended south to the city limits and eventually included building an overpass for rail traffic at Seventy-fifth Street. This view, looking north on Western Avenue from Seventy-fifth Street, shows a steam shovel preparing trenches for sewer piping. (Courtesy Ridge Historical Society.)

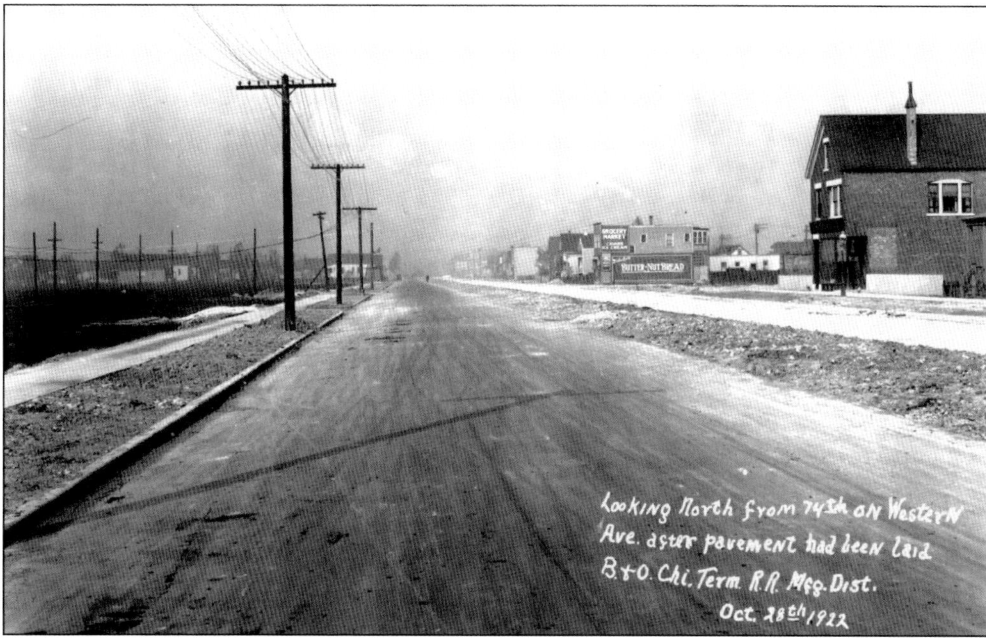

By October 1922, much of the sewer, curb, and paving work had been completed in Chicago on Western Avenue, the Dixie Highway. This view shows Western Avenue looking north of the Baltimore and Ohio Railroad tracks at Seventy-fourth Street. Although some of the buildings pictured still exist, the streetscape in this area is certainly much different today. (Courtesy Ridge Historical Society.)

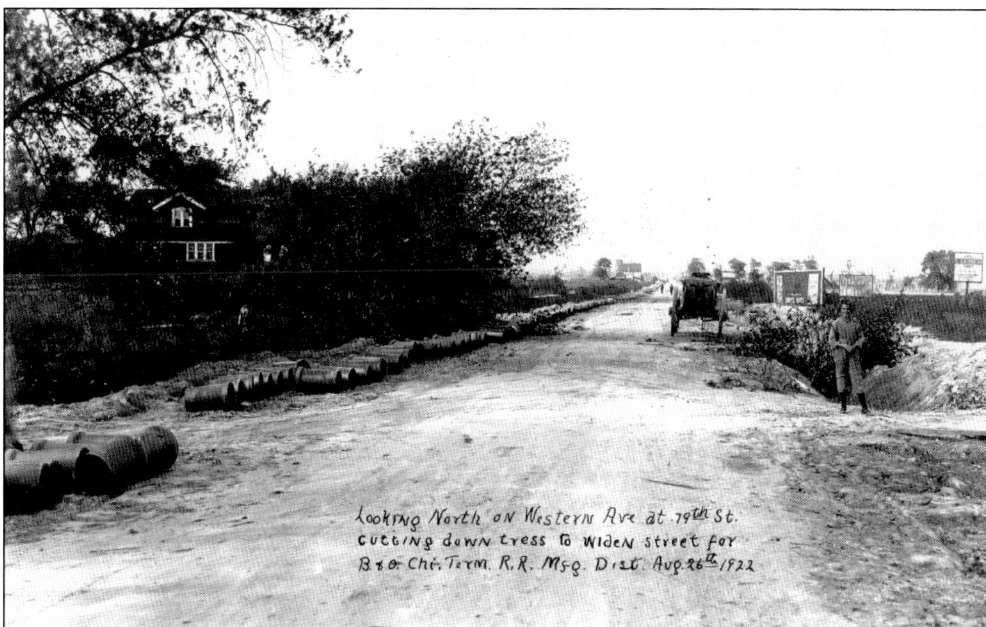

Even in Chicago, the Dixie Highway, Western Avenue, had a somewhat rural appearance in areas. This photograph, documenting progress on the widening and paving project, was taken in August 1922 and looks north on the highway at Seventy-ninth Street. A young girl proudly poses for the camera at right. (Courtesy Ridge Historical Society.)

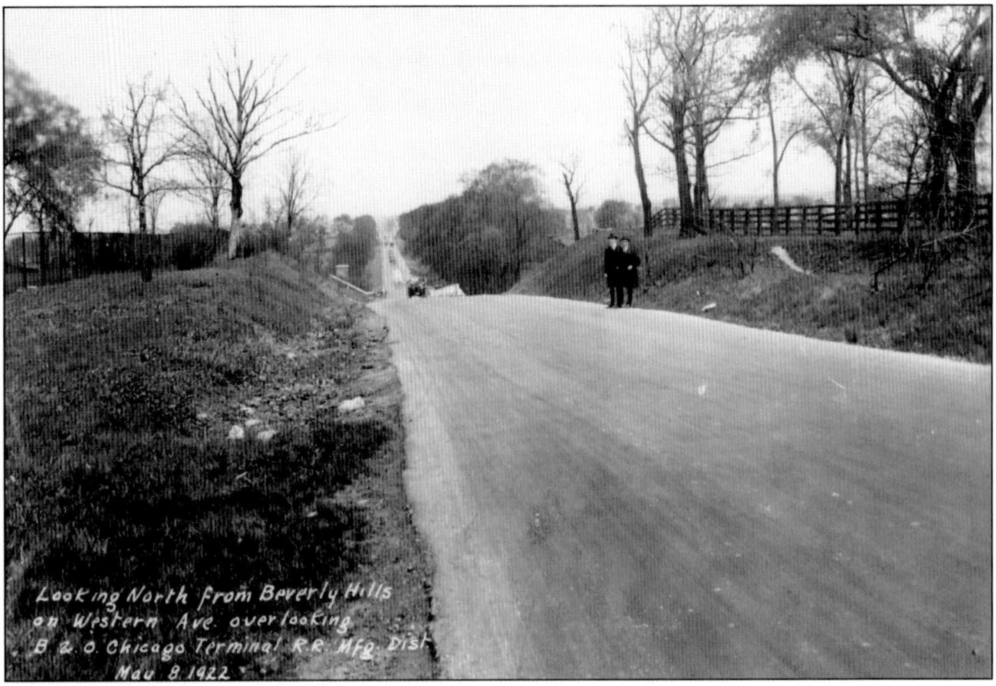

A part of the Western Avenue improvement project in Chicago near the Beverly Country Club included leveling the hill between Eighty-seventh and Ninety-first Streets to allow traffic to climb the grade more easily. In this view, highway engineers examine the hill and pavement before work on the project begins in May 1922. (Courtesy Ridge Historical Society.)

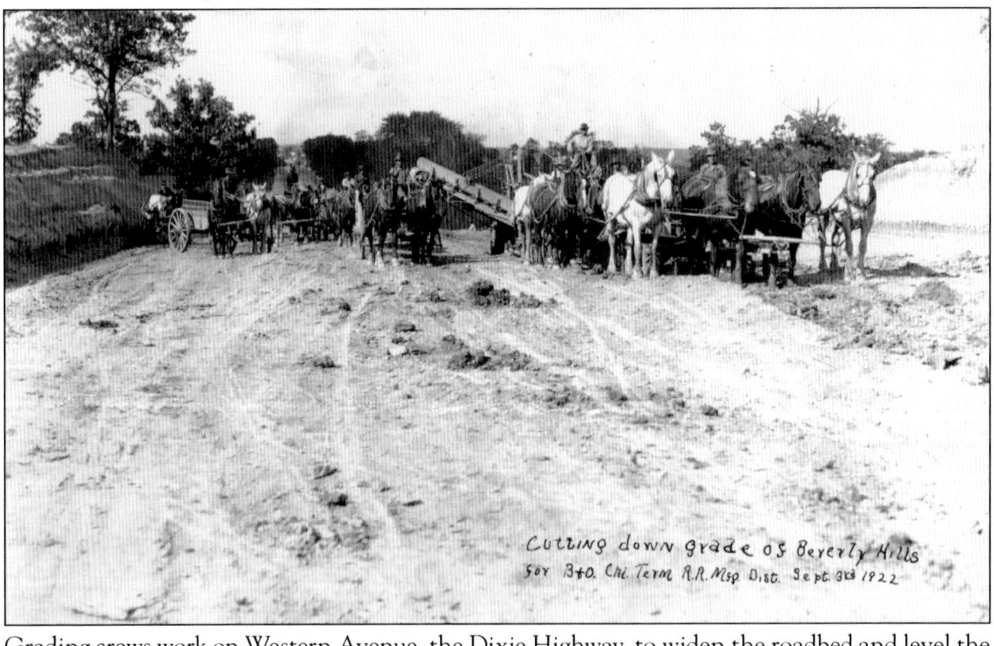

Grading crews work on Western Avenue, the Dixie Highway, to widen the roadbed and level the grade adjacent to the Beverly Country Club north of Ninety-first Street in Chicago. Although the work was done well into the motorized age, the contractor was still using teams of horses to complete this demanding aspect of the project. (Courtesy Ridge Historical Society.)

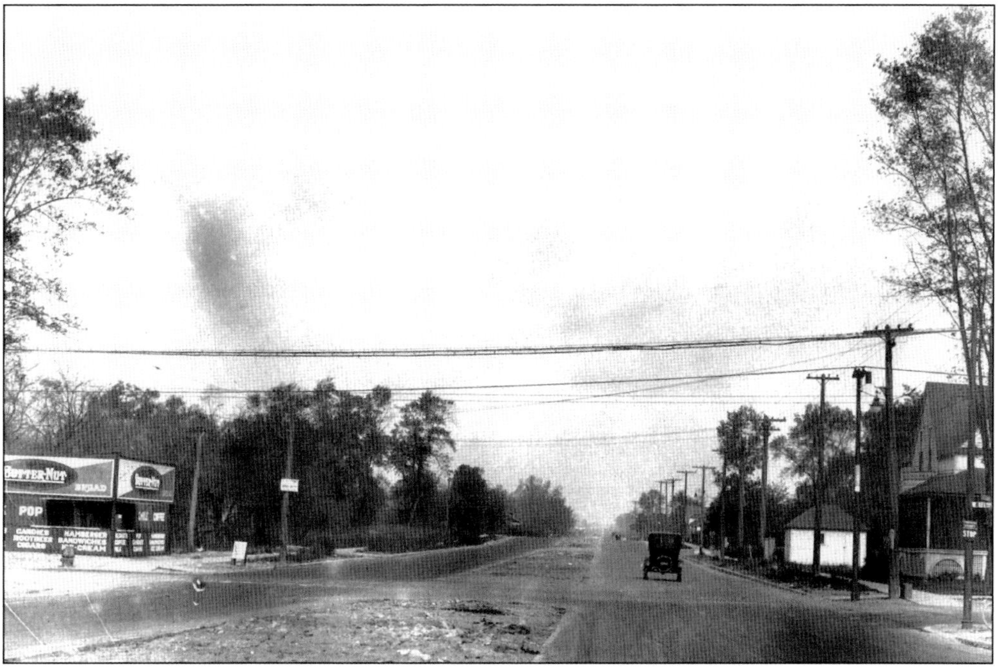

Traffic flows freely along Western Avenue, the Dixie Highway, through the intersection of 111th Street in Chicago. Completion of the Western Avenue widening and pavement improvement project aided the tremendous growth that the Beverly and Mount Greenwood neighborhoods of Chicago experienced in the 1920s. (Courtesy Ridge Historical Society.)

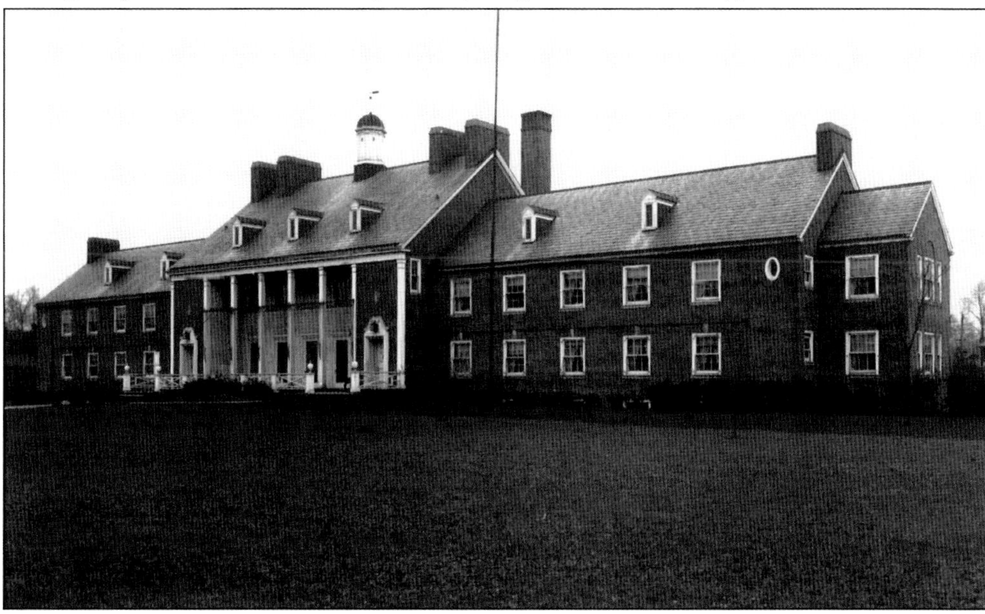

The Oakhaven Old People's Home in Chicago opened in 1924. The original building was constructed facing Western Avenue, the Dixie Highway, on a parcel between 112th Place and 113th Place and had accommodations for 43 residents. In 1929, the home merged with the Washington and Jane Smith Trust and assumed that name. Today Smith Village continues to provide senior living in newly completed buildings on this site. (Courtesy Ken Jellema.)

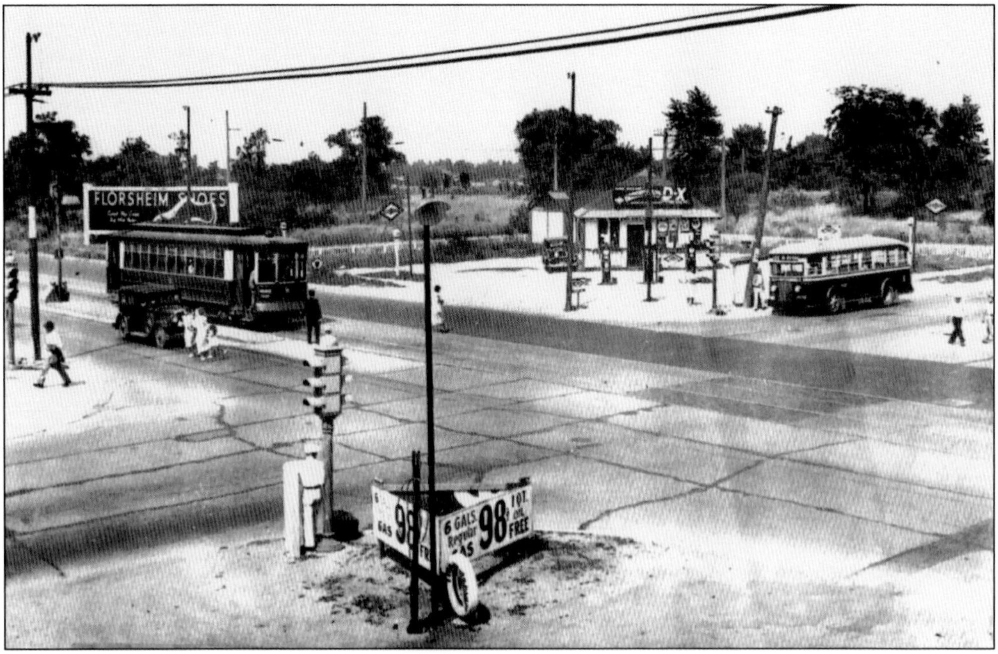

The west side of Western Avenue, the Dixie Highway, serves as the eastern border of Evergreen Park. This photograph from 1935 illustrates the intersection of Ninety-fifth Street and Western Avenue. On the right, a suburban transit bus is headed west, bound for Oak Lawn, while a "red rocket" Chicago Surface Lines streetcar heads south on Western Avenue. The sign in the foreground advertises six gallons of gasoline for 98¢. (Courtesy Evergreen Park Historical Commission.)

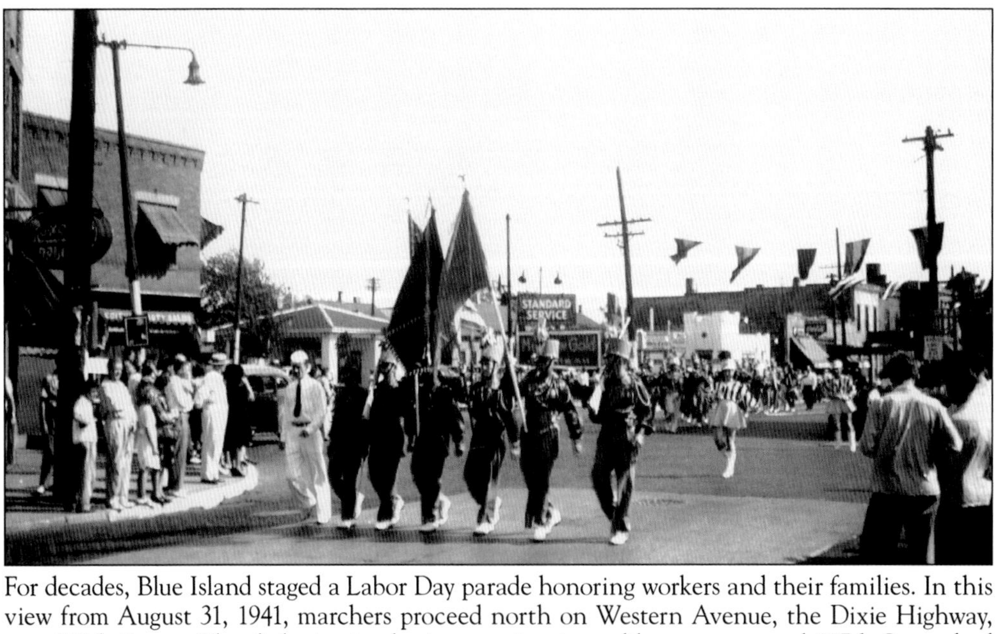

For decades, Blue Island staged a Labor Day parade honoring workers and their families. In this view from August 31, 1941, marchers proceed north on Western Avenue, the Dixie Highway, past 127th Street. The slight jog in the intersection is readily apparent, and 127th Street had not yet been widened when this photograph was taken. The original White Castle restaurant is visible in the distance. (Courtesy Blue Island Historical Society.)

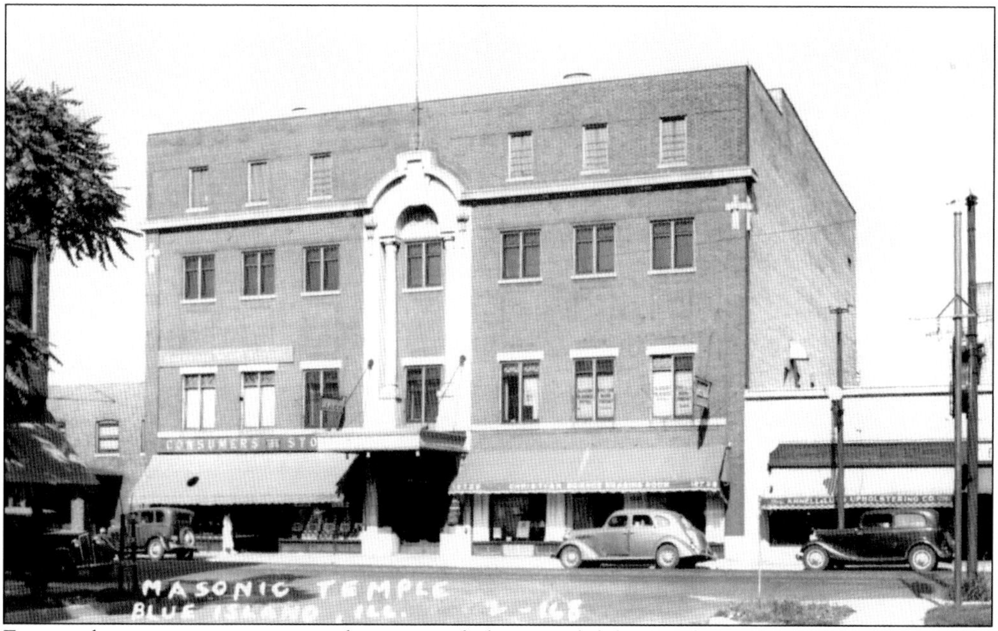

Fraternal organizations were a big part of the social life in Blue Island. Several of these organizations even built their own halls. The Masonic temple was built in 1921 on the east side of Western Avenue, the Dixie Highway, at Oak Street. The Consumers Store and the Christian Science Reading Room occupied the ground level, and the second floor was rented for office space. The Masons used the upper floors for their activities. (Courtesy Richard Reuss.)

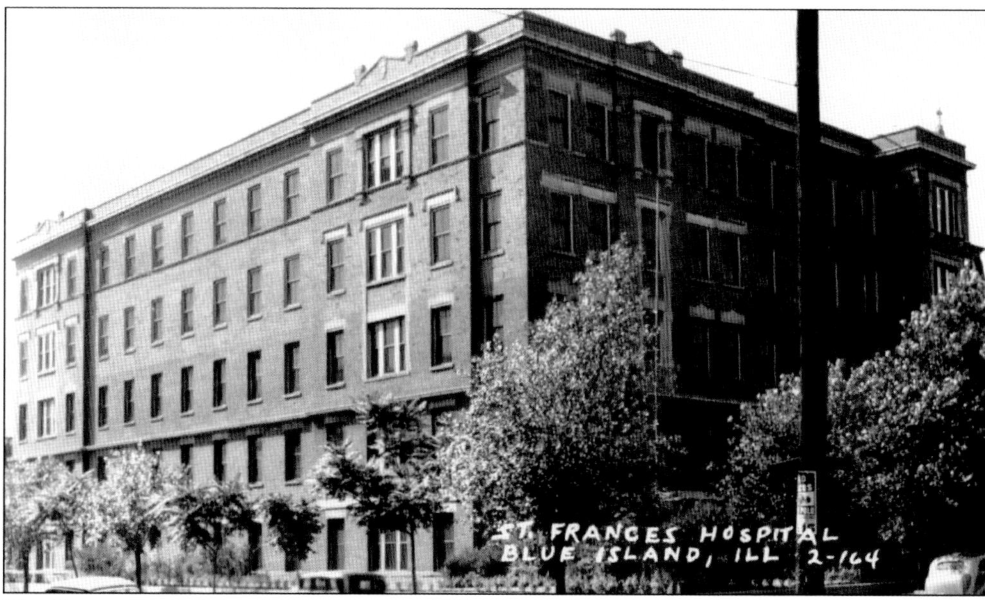

The Franciscan Sisters of St. Mary, a Catholic order of nuns, established St. Francis Hospital in Blue Island on Gregory Street, just east of the Dixie Highway, in 1905. The hospital was housed in the Uhlich mansion until 1916 when a modern five-story building with 100 rooms was dedicated that Memorial Day. The hospital has grown over the years, and in 2008, after 103 years, the sisters sold the hospital facility. It remains in operation under the name MetroSouth Medical Center. (Courtesy Richard Reuss.)

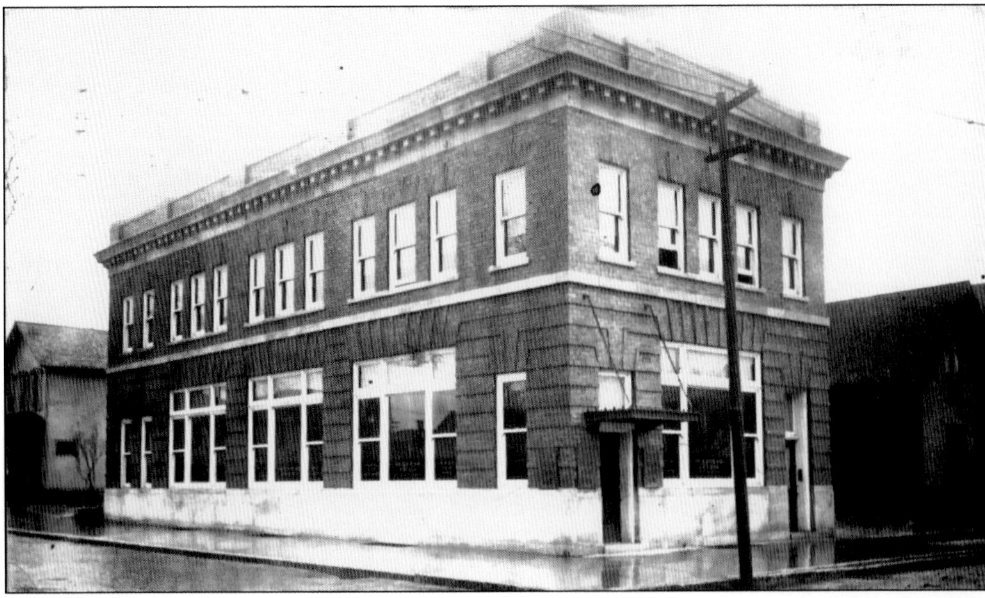

In addition to being an industrial center, Blue Island was also a financial center in the south suburbs. The Blue Island State Bank organized in 1912 and opened for business in this building on the northwest corner of Western Avenue, the Dixie Highway, and High Street. In 1928, the bank merged with the Blue Island Trust and Savings Bank, which was established two years earlier. Today a Laundromat occupies the first floor of the building. (Courtesy Richard Reuss.)

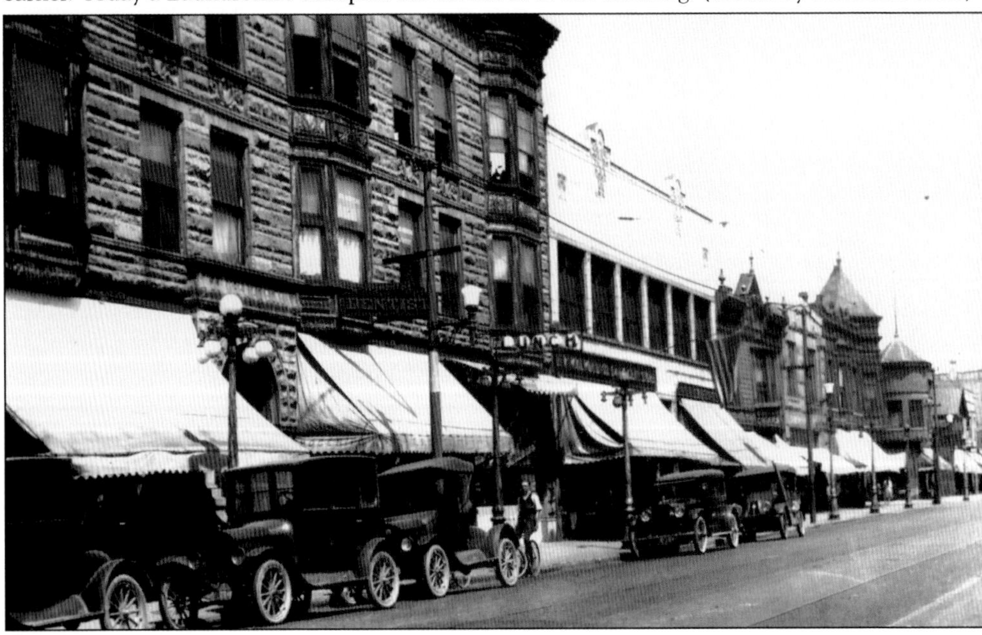

In Blue Island, Western Avenue has always been a popular retail strip. The Woolworth's store was a mainstay on the street for decades and occupied a white terra-cotta brick building on the west side of Western Avenue between New and Vermont Streets, pictured in the center of this photograph. The Elks Club maintained its quarters on the second floor of this building, which is currently home to the Three Sisters Antique Mall. The Montag building, at left, was torn down in the 1940s. (Courtesy Richard Reuss.)

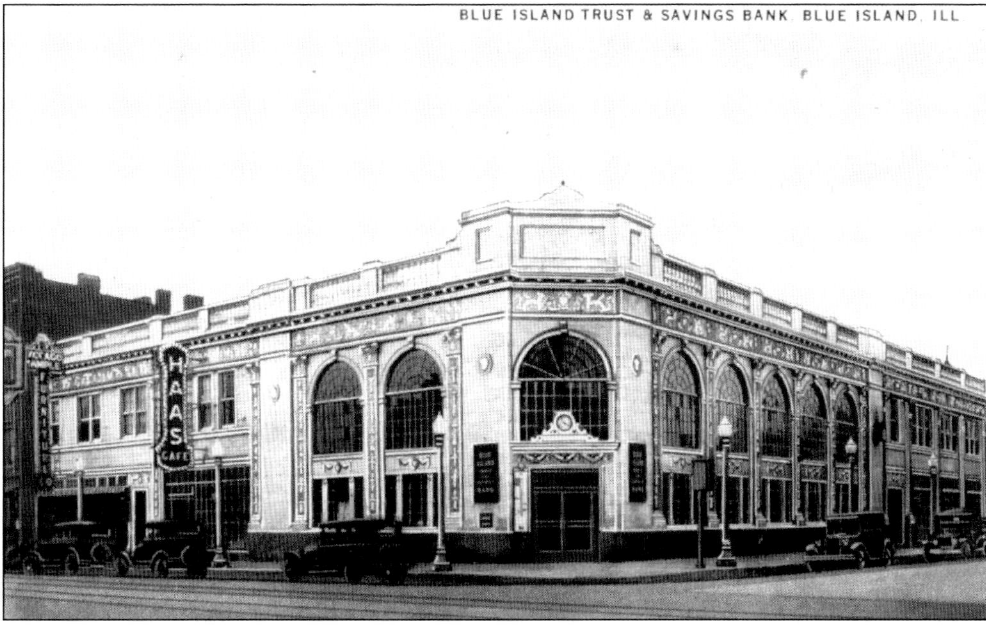

Many new buildings were constructed in Blue Island along the Dixie Highway, Western Avenue, in the Roaring Twenties. The Blue Island Trust and Savings Bank built this beautiful terra-cotta brick structure with many decorative elements inside and out on the southwest corner of Vermont Street and Western Avenue. The bank opened for business in March 1926 and was reorganized three years later as the State Bank of Blue Island. The bank went into receivership in August 1933. (Courtesy Richard Reuss.)

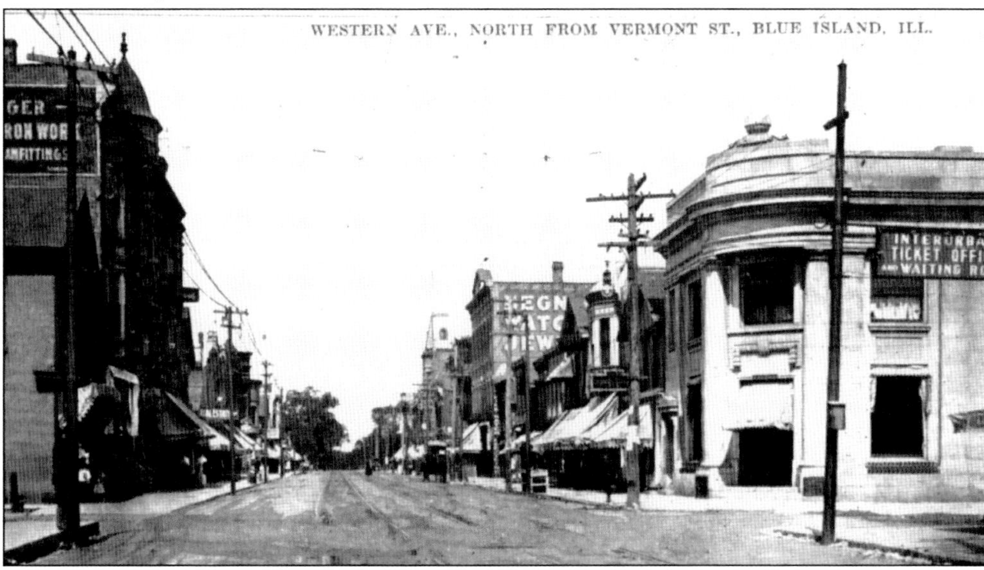

This view of Western Avenue in Blue Island was taken after the beginning of the 20th century. The Chicago and Interurban Traction Company's streetcar tracks are visible along with a sign for the company's ticket office and waiting room at right. The interurban ceased operations in 1927. Also at right is the headquarters of the First National Bank of Blue Island, which was organized in 1896. The bank, now known as Great Lakes Bank, still operates out of these quarters, which have been enlarged over the years. (Courtesy Richard Reuss.)

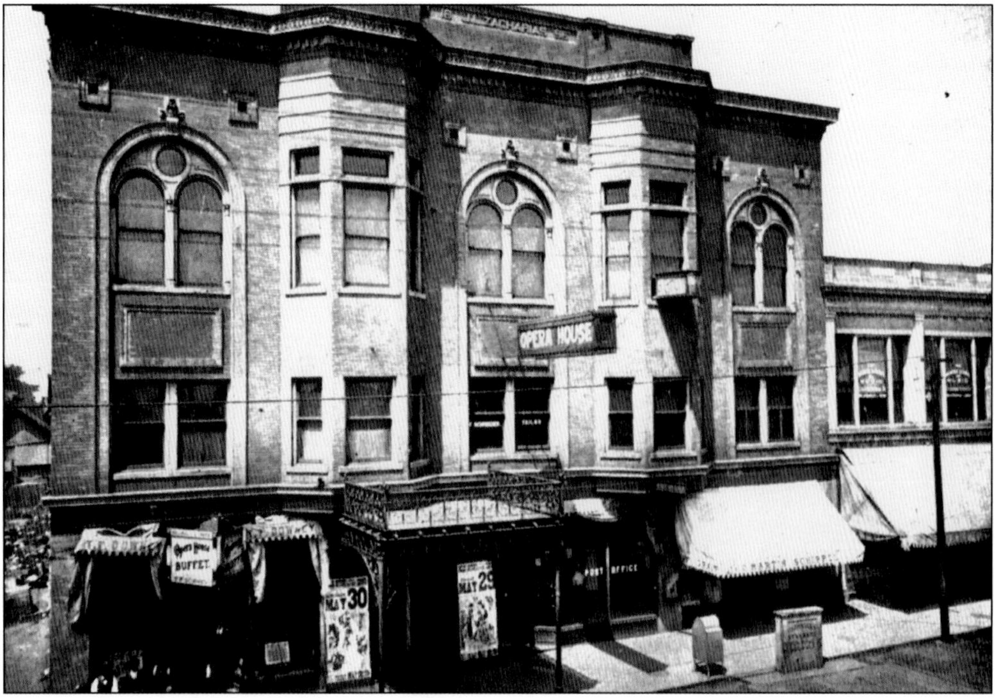

Western Avenue, the Dixie Highway, in Blue Island has also been a destination for entertainment. The Blue Island Opera House at 13114 Western Avenue was built after the original building on the site was destroyed by the great Blue Island fire in 1896. In later years, the Grand Theatre was located in the building, which has been restored and designated a local landmark. (Courtesy Richard Reuss.)

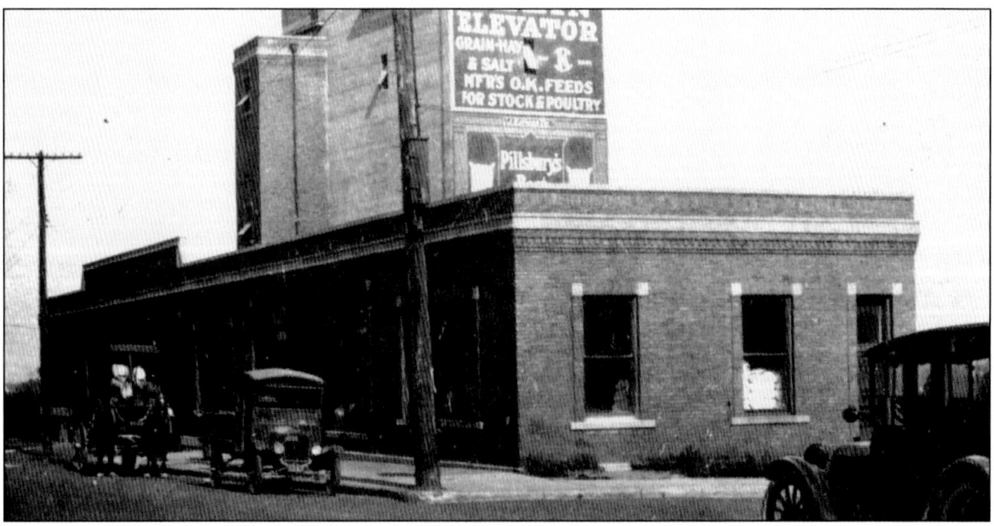

The Klein Elevator in Blue Island was a landmark visible for a great distance to motorists on the Dixie Highway. Located a block east of Western Avenue on Vermont Street, the original structure burned down in April 1914, and a more fire-resistant replacement was started almost immediately. The rebuilt elevator, a reminder of the area's agricultural roots, was torn down in the mid-1960s to make way for the realignment of northbound Western Avenue traffic onto Gregory Street. (Courtesy Richard Reuss.)

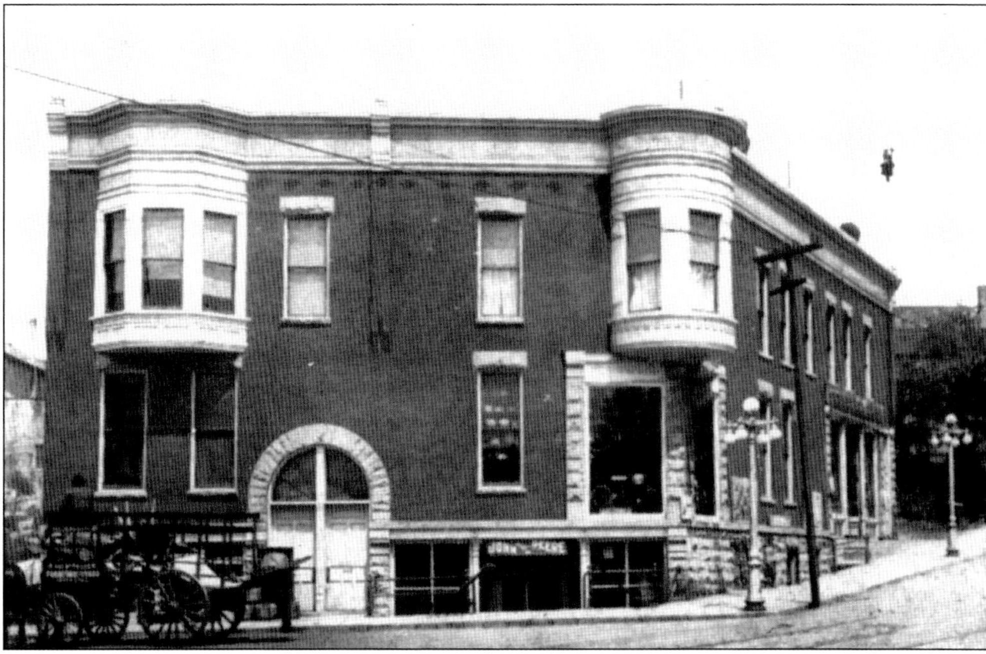

August Schreiber emigrated from Germany to Blue Island in the late 1840s and worked as a tinsmith briefly before opening a hardware store on Western Avenue at James Street. Successive generations of his family ran the store selling implements and a variety of hardware and plumbing supplies. This building, which dates to 1879, still stands, and an art gallery was the most recent occupant of the lower level. (Courtesy Richard Reuss.)

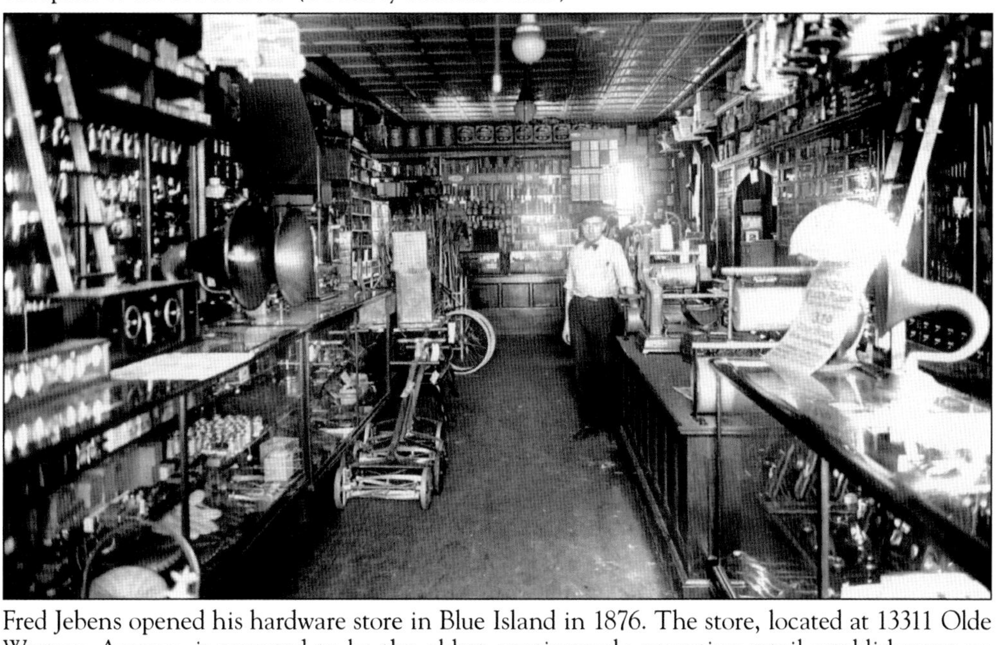

Fred Jebens opened his hardware store in Blue Island in 1876. The store, located at 13311 Olde Western Avenue, is reported to be the oldest continuously operating retail establishment on the Dixie Highway in Illinois. The old-fashioned hardware store has expanded its business by serving contractors and thrives by offering customers personalized service. (Courtesy Jebens Hardware.)

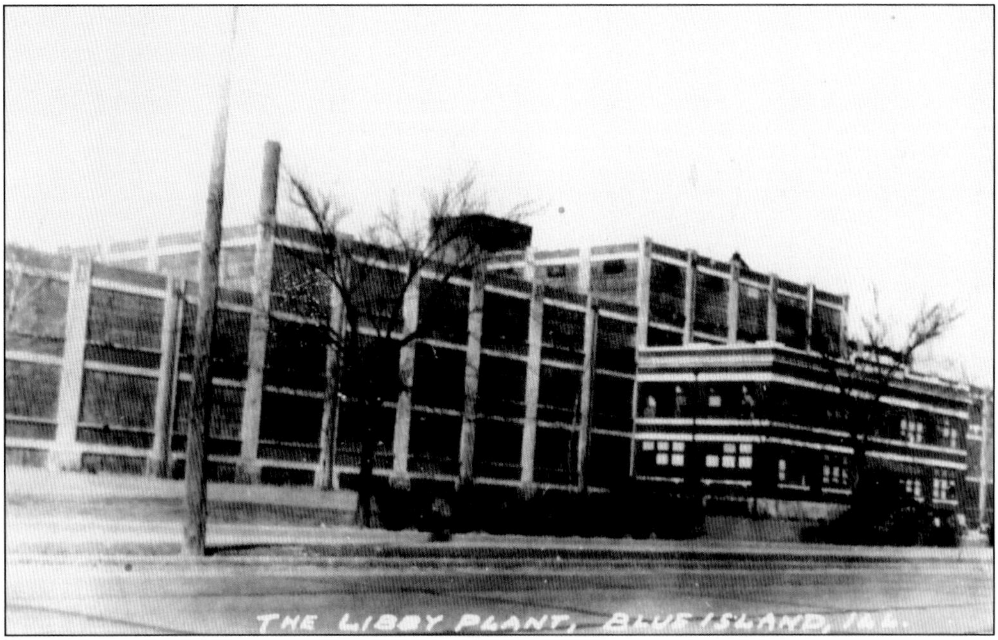

Libby, McNeil and Libby constructed this modern canning plant in Blue Island on 24 acres adjoining the Dixie Highway in 1918. Considered to be of the most modern design, the company canned a wide variety of products in the facility including olives, applesauce, pork and beans, ketchup, jellies, mustard, pickles, relish, tomato sauce, and juice. The building still stands and is used by a number of small companies as an intermodal terminal. (Courtesy Richard Reuss.)

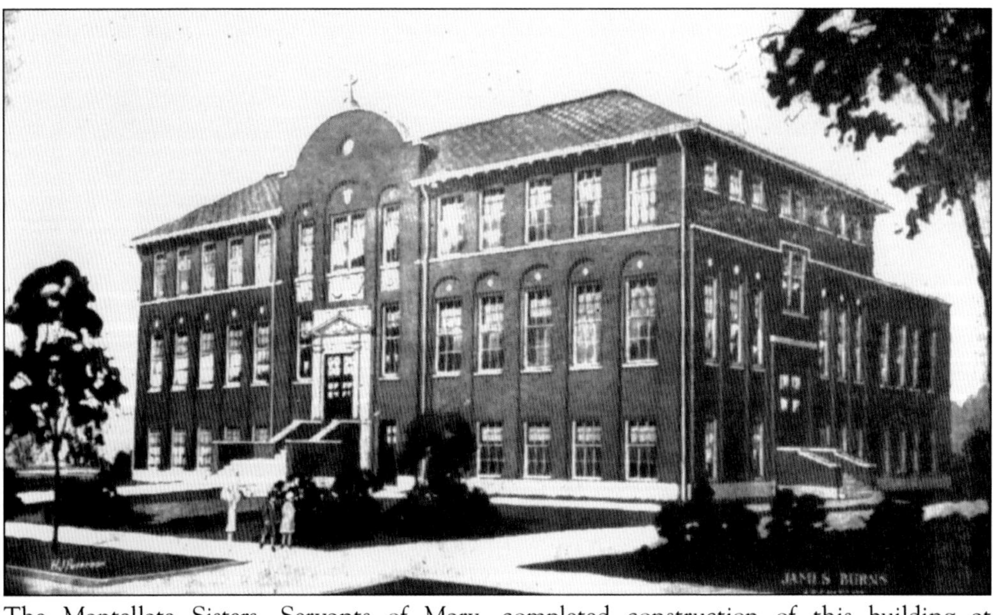

The Mantellate Sisters, Servants of Mary, completed construction of this building at 13811 Western Avenue, the Dixie Highway, in Blue Island, in 1928. The sisters established Mother of Sorrows Institute, an elementary school, which was furnished with modern classrooms and dormitory facilities for boys and girls. By 1954, the sisters were operating an all-girls high school at this location. The school closed in 1989. (Courtesy Blue Island Public Library.)

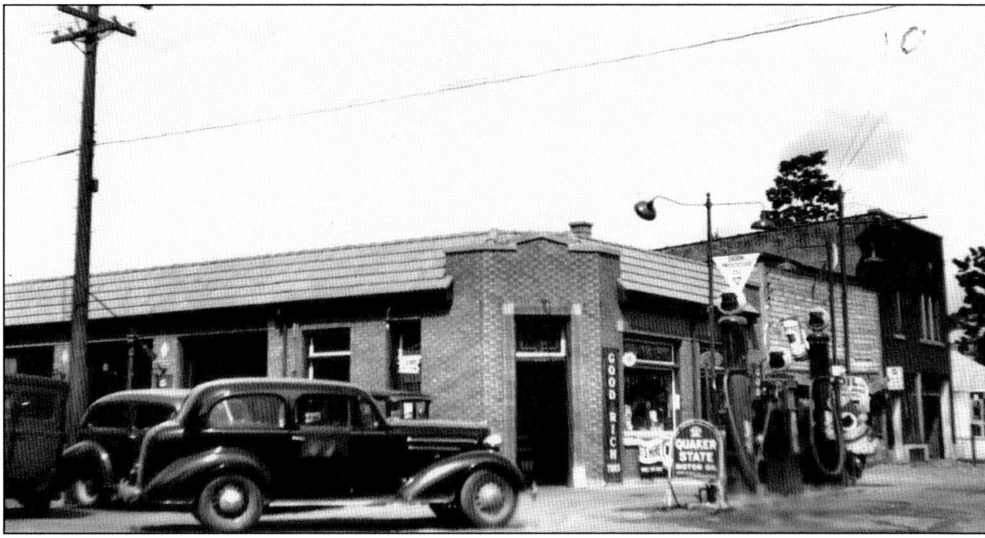

Bauer's Super Service was located on the northwest corner of 139th Street and Western Avenue in Blue Island. The station sold Phillips 66 gasoline, Quaker State motor oils, and Goodrich tires. Bauer's did a good business as the three garage bays and the cars waiting to be serviced attest. Vargas Mufflers and Brakes occupies the site today. (Courtesy Vargas Mufflers and Brakes.)

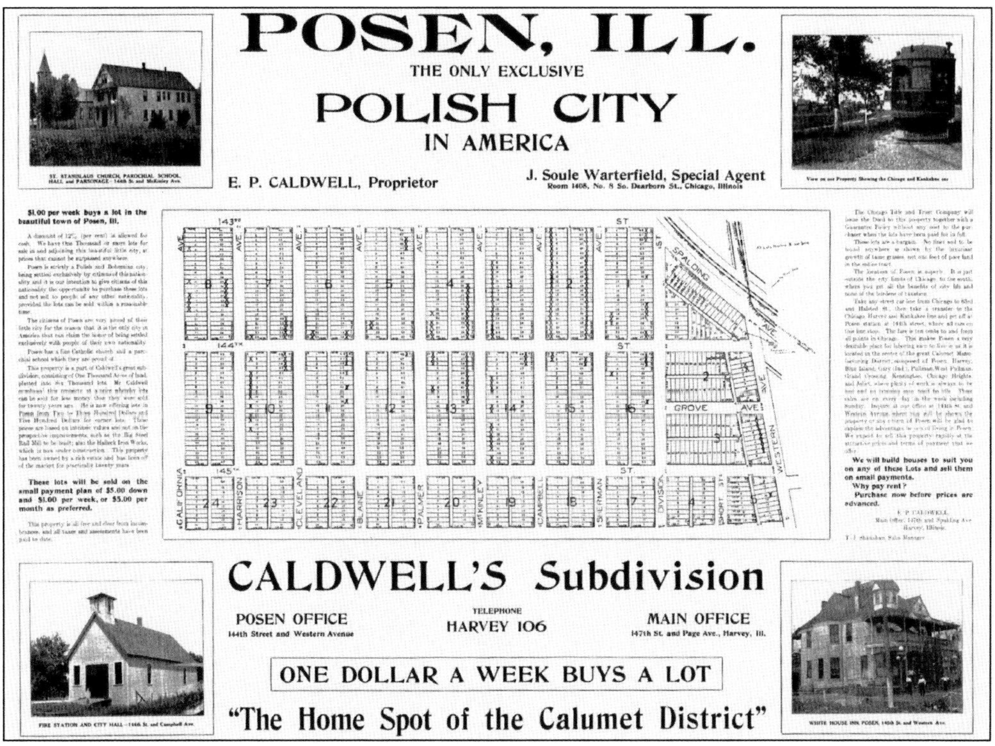

The Dixie Highway forms the eastern border of the village of Posen between 147th and 159th Streets. The village was incorporated in 1900, and the name of the community was selected because many residents immigrated from in and around Poznan (Posen), Poland. Early land subdivider E. P. Caldwell successfully marketed his real estate in the community as being in "the only exclusive Polish city in America." (Courtesy Village of Posen.)

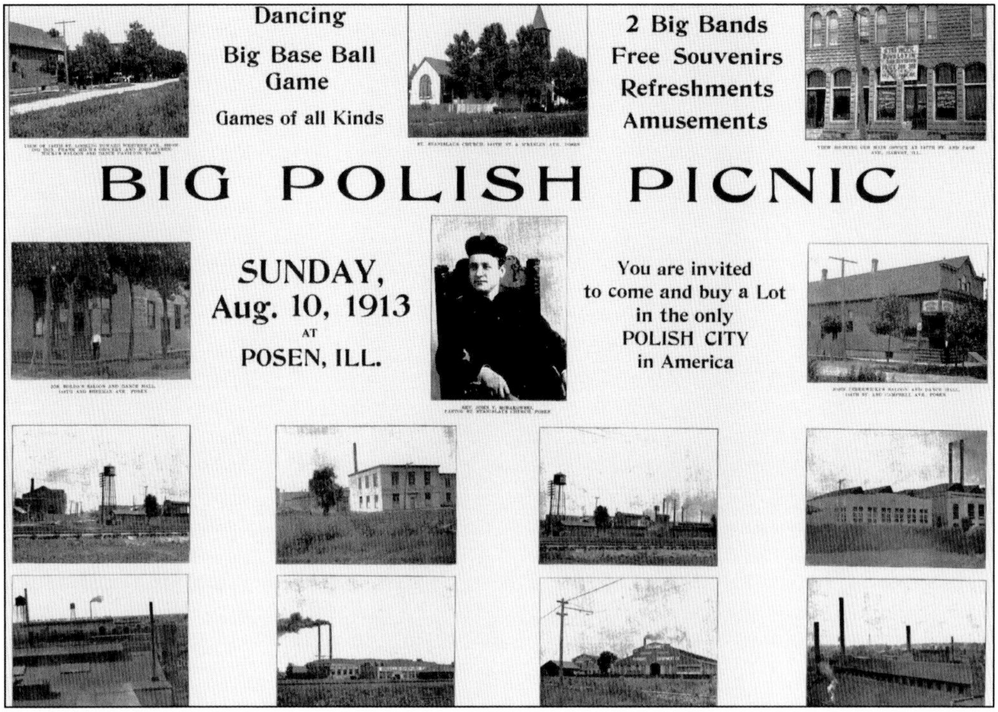

E. P. Caldwell used creative means to market his subdivision in Posen. In addition to handbills and flyers, Caldwell also sponsored a picnic in the community meant to show prospective buyers the quality of life they could enjoy in the village, which was within walking distance of many major employers in neighboring Harvey. Caldwell also prominently featured images of the town's Catholic church and parish priest, important institutions to Polish immigrants. (Courtesy Village of Posen.)

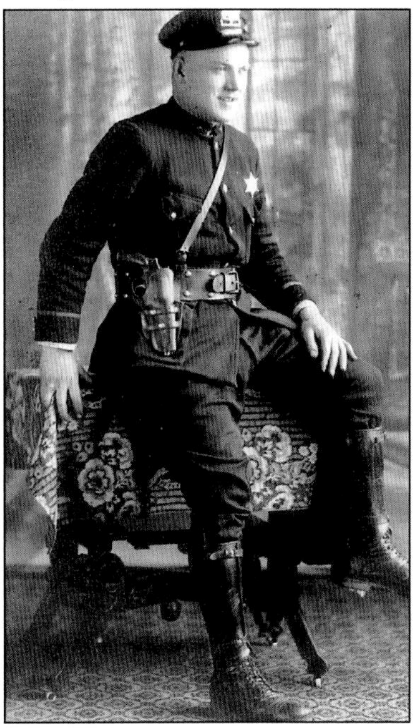

Officer Tom Leuther presented an imposing appearance on patrol with his motorcycle on the streets of Dixmoor. Leuther was a member of the village's police force in the 1930s and no doubt nabbed a few speeders on the Dixie Highway. The highway formed the western border of this small village south of Blue Island between 139th and 147th Streets. (Courtesy Geraldine Leuther Justice.)

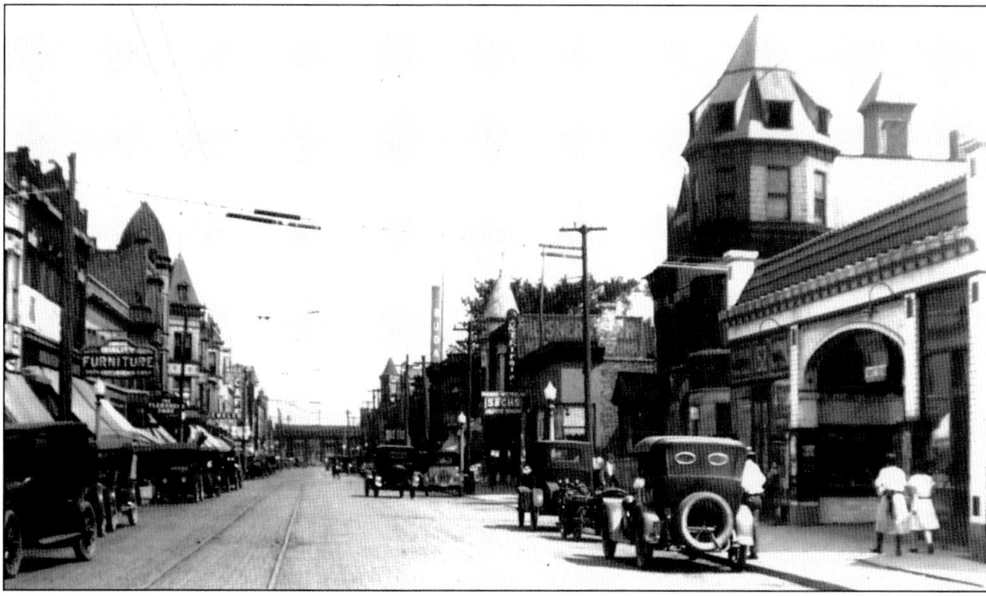

The Dixie Highway travels for a mile, from 147th to 159th Streets, along the western border of Harvey. During the early years of the roadway, this stretch was sparsely populated, and few businesses were located there. Just east of the highway on 154th Street, however, the downtown business district offered travelers a wide range of goods and services to satisfy their needs. (Courtesy Debbie Lundmark.)

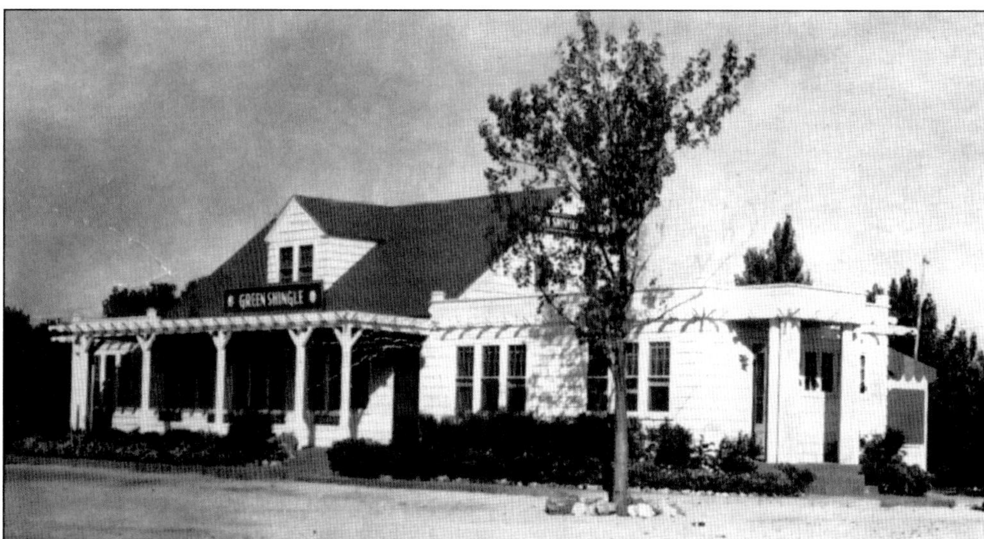

The Green Shingle, namesake restaurant of owners Art and Mabel Green, was a Harvey landmark for decades at the corner of 159th Street, U.S. Route 6, and the Dixie Highway. The restaurant was known for its chicken, steak, and fish dinners. During summer months, patrons could enjoy dining and dancing in beautifully landscaped outdoor gardens. The restaurant closed in the early 1980s, and a Shell gasoline station occupies the site. (Courtesy Debbie Lundmark.)

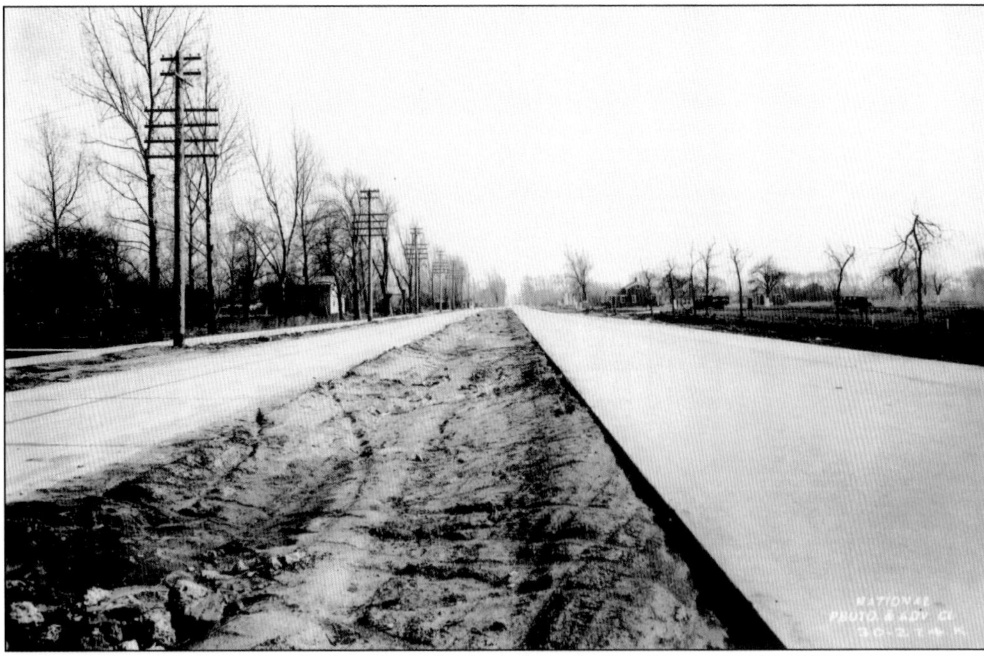

New concrete pavement for the Dixie Highway south of 167th Street in Hazel Crest is visible on the right. The new pavement bypassed the old slab used for the highway at left. The old slab is now called Robey Avenue. (Courtesy Homewood Historical Society.)

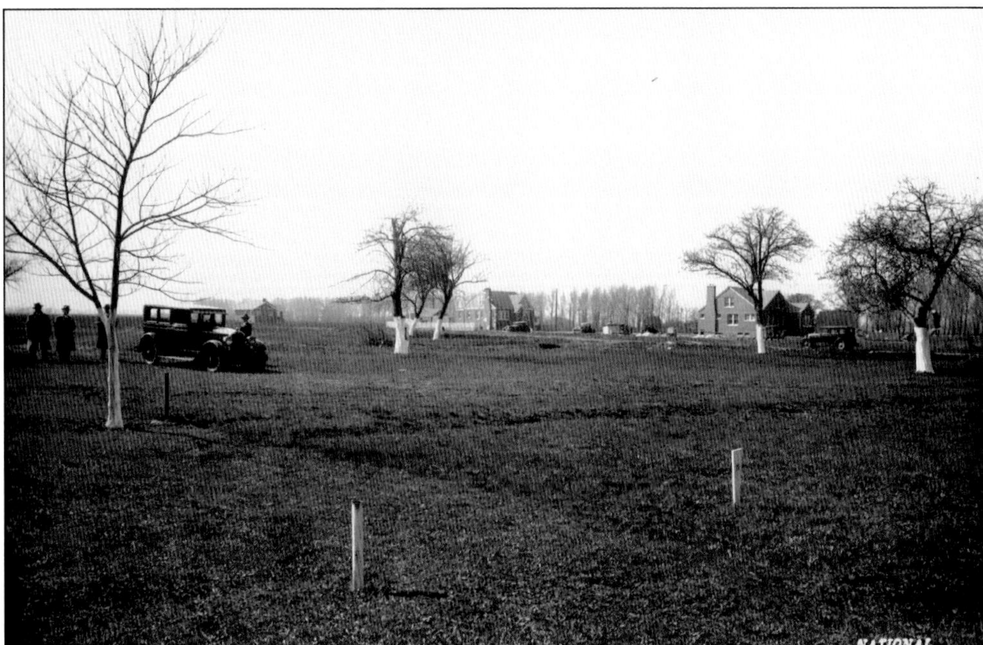

Construction on the Hazel Crest Park subdivision was just beginning when this photograph looking northwest from Dixie Highway to what is now 168th Street and Shea Avenue was taken in 1930. The onset of the Depression halted any further development on this project for years. (Courtesy Homewood Historical Society.)

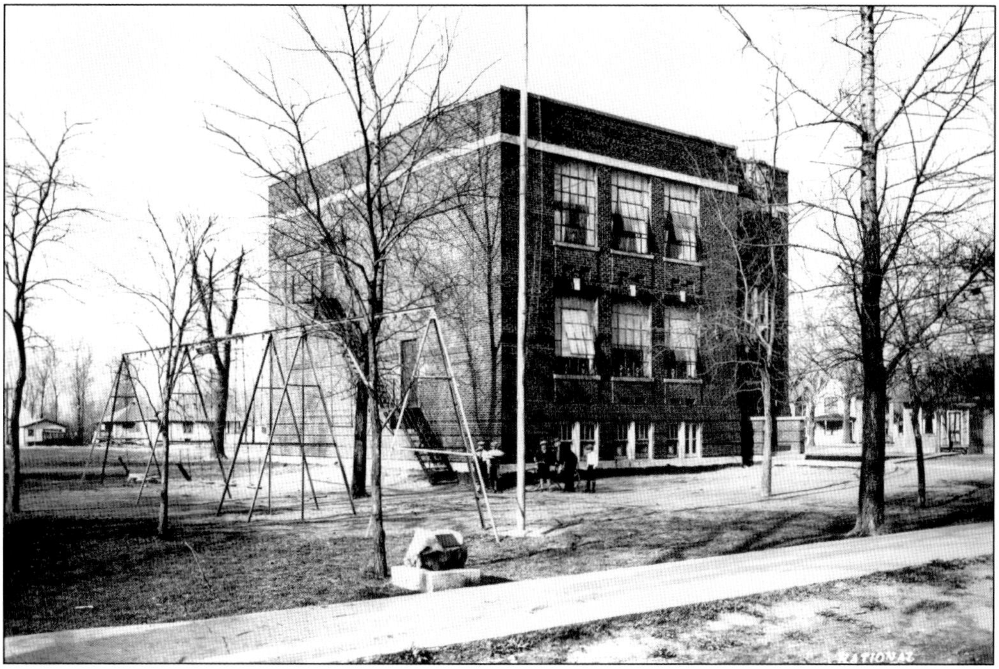

A group of youngsters play in front of Roosevelt School located just east of the Dixie Highway at 170th Street. The Hazel Crest school has been enlarged over the years and is now named Warren Palm, in honor of the school district's first school board president. (Courtesy Homewood Historical Society.)

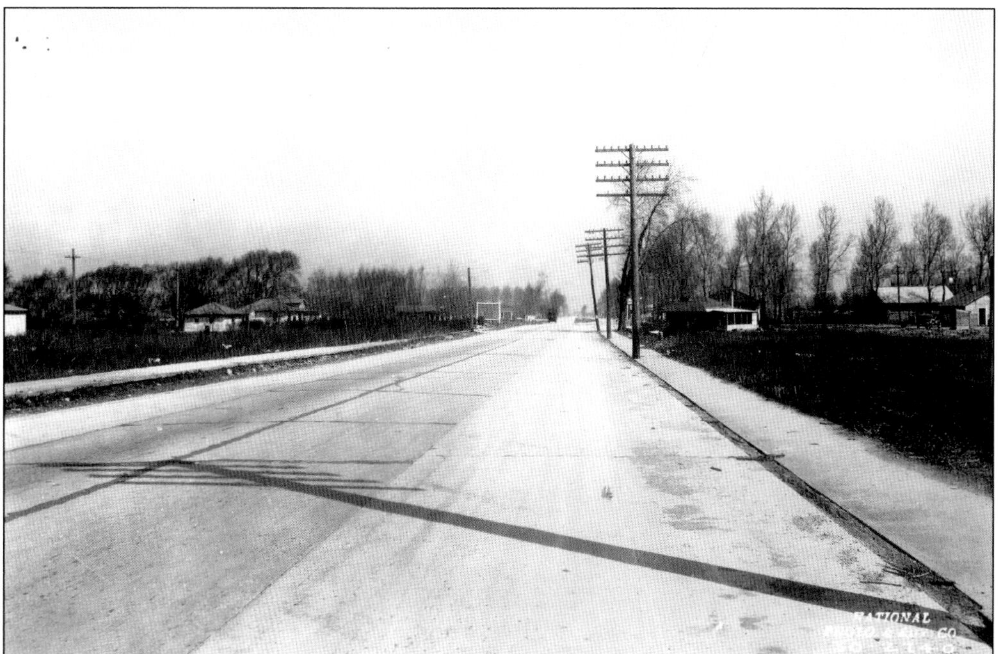

New concrete pavement on the Dixie Highway is visible in this view from 1930 looking north from 170th Street. The undeveloped nature of Hazel Crest at this time is readily apparent. (Courtesy Homewood Historical Society.)

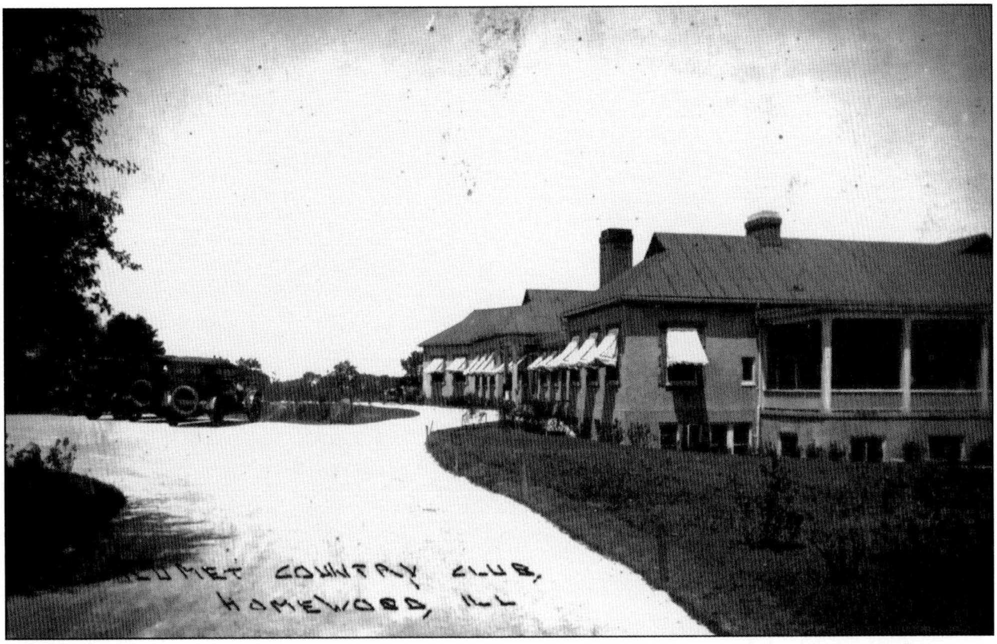

Homewood has been a golf community for over a century. Just west of the Dixie Highway on 175th Street, Calumet Country Club relocated to the village in 1917. The club originally was established in 1901 on acreage near Ninety-fifth Street and Cottage Grove Avenue in Chicago, now the site of Chicago State University. (Courtesy Homewood Historical Society.)

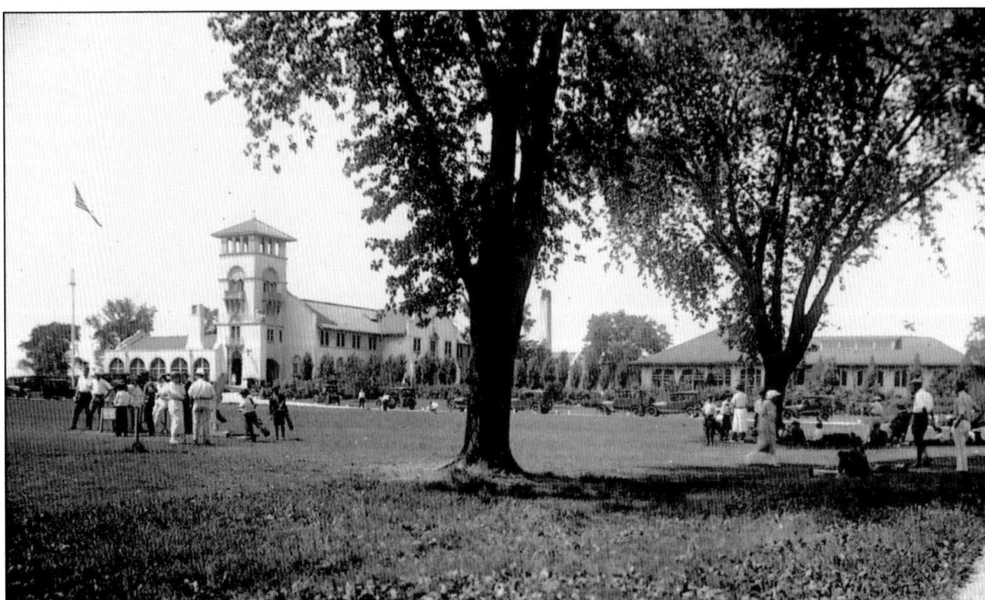

Ravisloe Country Club was another golf course located near downtown Homewood just west of the Dixie Highway. The club was established in the village in 1901 and its Spanish Mission style clubhouse is a south suburban landmark. Originally members arrived by train from Chicago for a day of golf, but as this photograph illustrates, most were coming by car in the 1920s. (Courtesy Homewood Historical Society.)

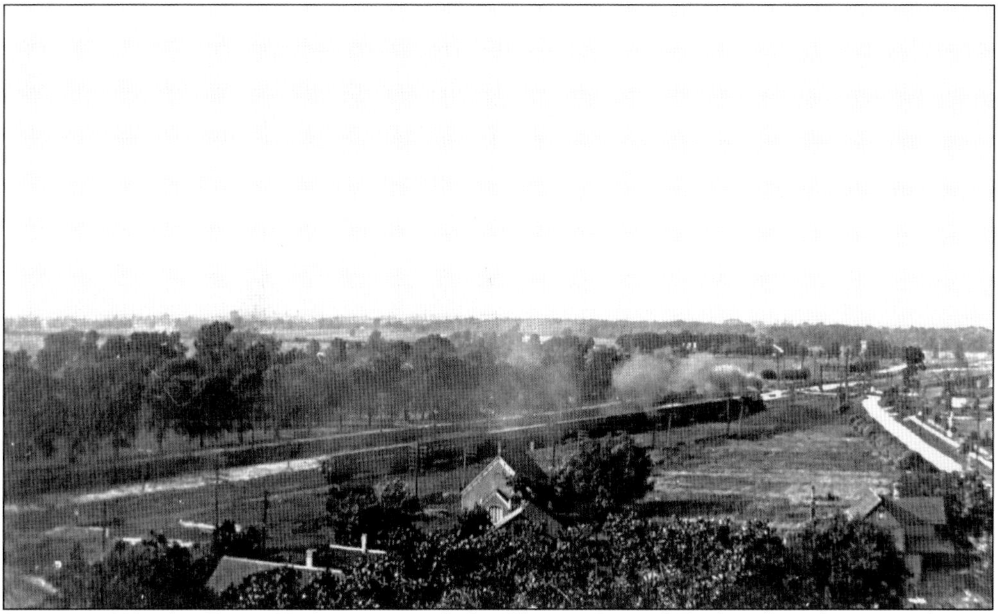

Prior to 1922, the Dixie Highway crossed the Illinois Central Railroad tracks on grade just north of downtown Homewood near what is now Maple Road. The photograph illustrates this dangerous crossing at right center. In 1919 alone, eight people were killed at this crossing, which was sadly known as "Death's Angle." (Courtesy Homewood Historical Society.)

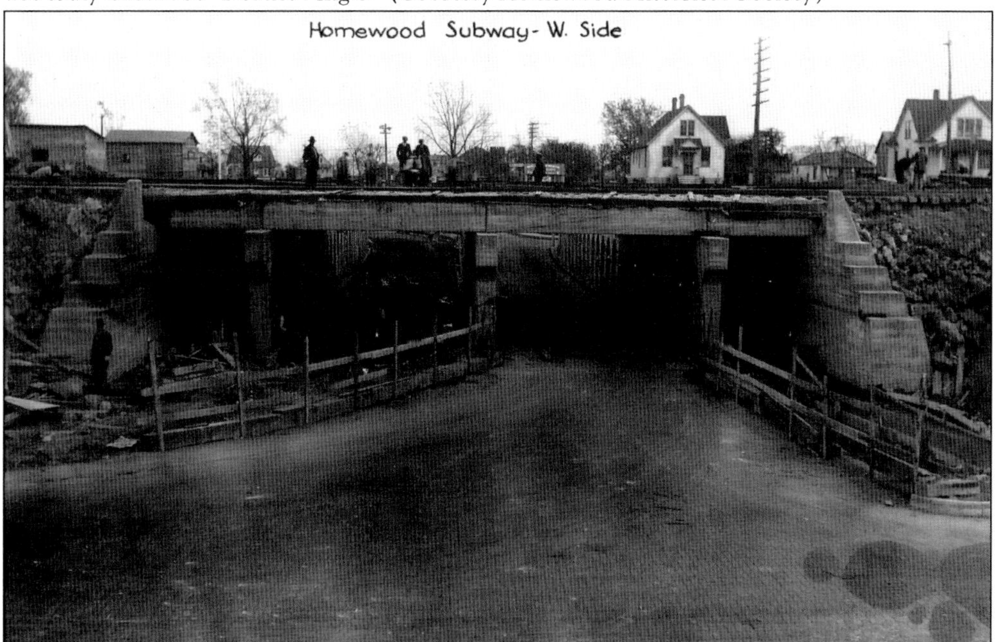

This October 1922 view looking east shows the Dixie Highway subway or viaduct nearing completion. The viaduct eliminated grade crossing accidents in Homewood but frequently caused another nuisance motorists had to deal with—flooding. Many times during heavy rains, the viaduct flooded, necessitating lengthy detours around it. Today larger storm sewers have dramatically reduced viaduct closures due to flooding. (Courtesy Homewood Historical Society.)

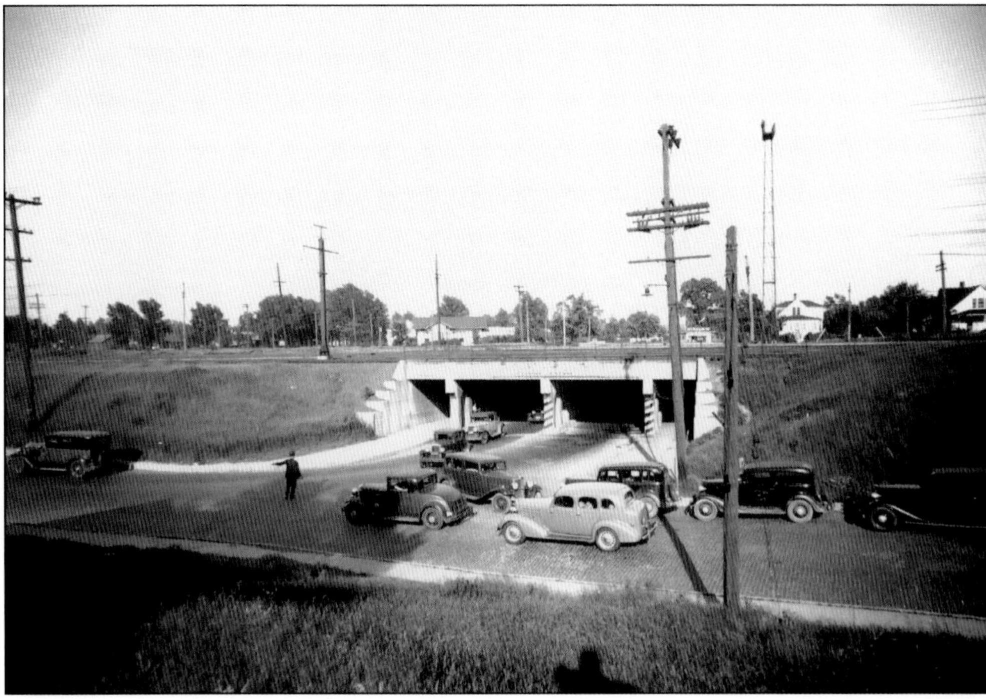

Traffic tie-ups at the Dixie Highway viaduct have also been an issue for motorists. For years, a Homewood policeman was stationed on the west side of the viaduct during the evening "rush hour" to direct traffic and ease congestion caused by an influx of cars from commuters leaving parking lots near the Homewood train station. (Courtesy Homewood Historical Society.)

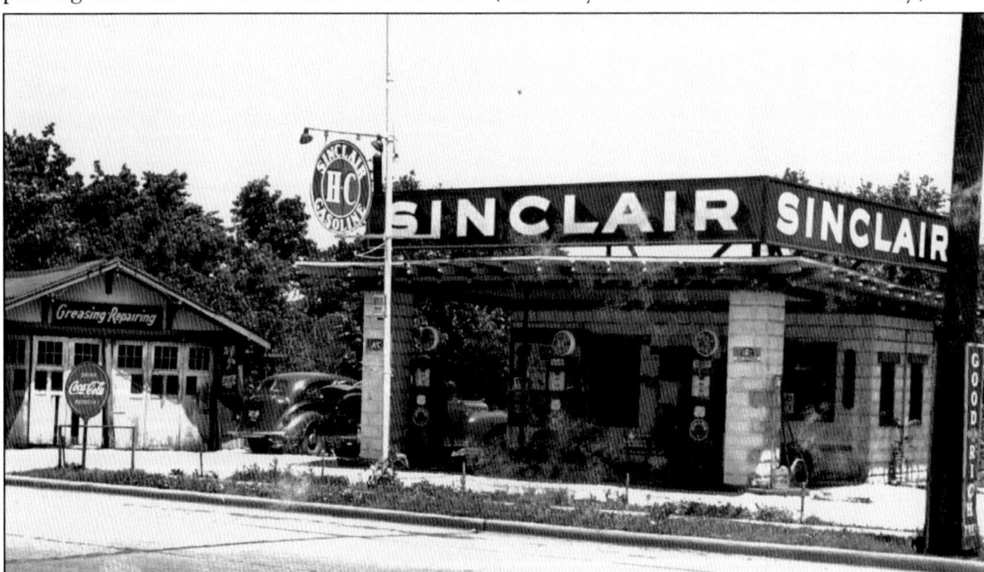

This Sinclair station at 17911 Dixie Highway in Homewood was owned by a number of operators before the Puhrmann family took it over. At first, repairs, oil changes, and grease jobs were done in the frame garage to the left. Later two service bays were added to the main building. A Clark Oil station subsequently occupied the site, which was torn down to allow for the construction of the Homewood Public Library. (Courtesy Homewood Historical Society.)

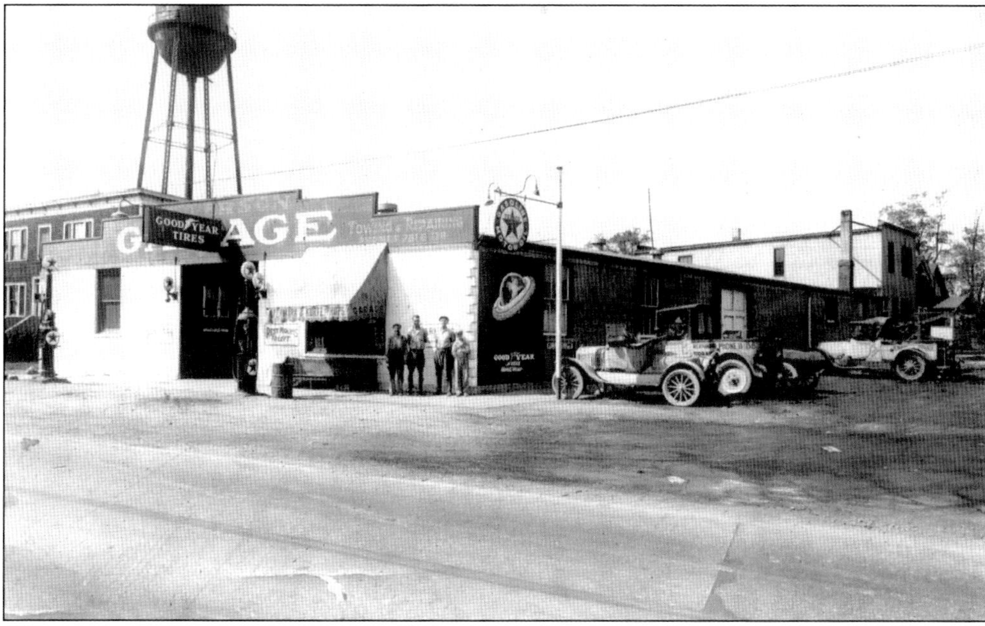

Ted Utermark and Jack Kurtz were the owners of the Severson Garage when this photograph was taken. The garage, built on the Dixie Highway in Homewood by Andrew Severson, Utermark's father-law, provided full automobile repairs and emergency towing service. The garage also sold Goodyear tires and Texaco gas and oils. The building was demolished in 1966, and the Homewood fire station now occupies the site. (Courtesy Homewood Historical Society.)

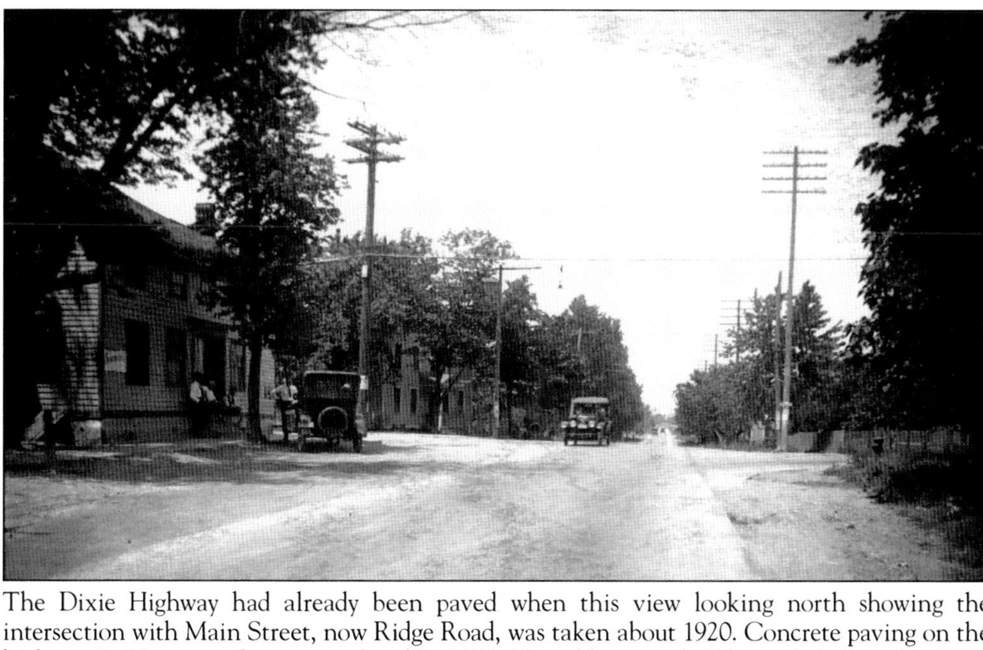

The Dixie Highway had already been paved when this view looking north showing the intersection with Main Street, now Ridge Road, was taken about 1920. Concrete paving on the highway in Homewood was completed in 1917. Two old tavern buildings dating to the 1850s occupied the west side of the road, while a residence and a blacksmith's shop rounded out the other buildings on the corner of this intersection. (Courtesy Homewood Historical Society.)

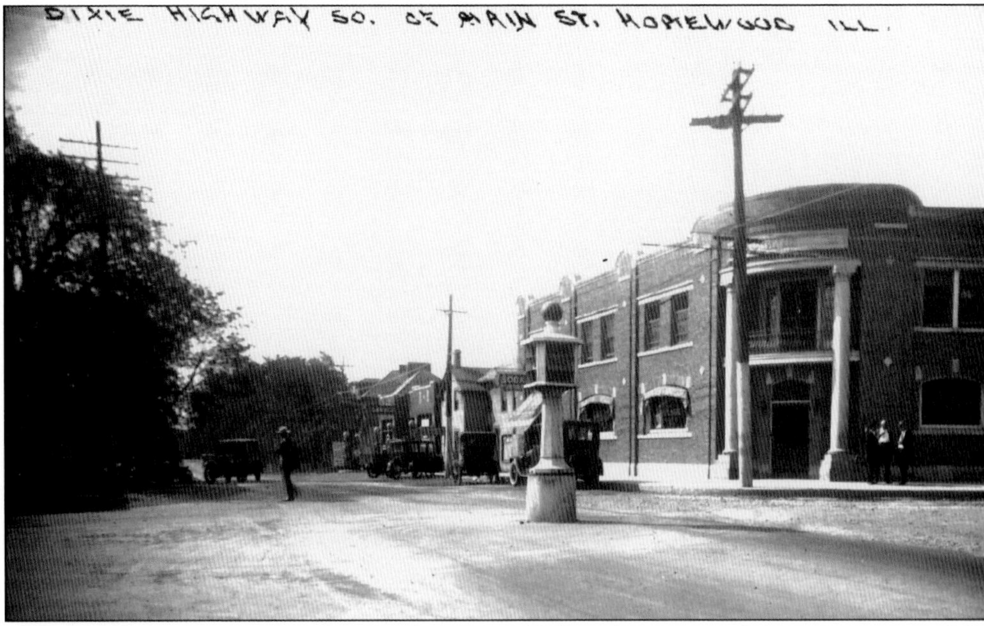

Homewood's first four-way traffic signal was installed in the center of the intersection of the Dixie Highway and Main Street (Ridge Road) in August 1917. The first signal was initially mounted on a wood post and was destroyed within months after being struck by a car. It was replaced by this signal mounted on a more substantial concrete base. The Cook County Trust and Savings Bank (right) opened for business in 1925, replacing the pioneer tavern that had previously occupied the site. (Courtesy Homewood Historical Society.)

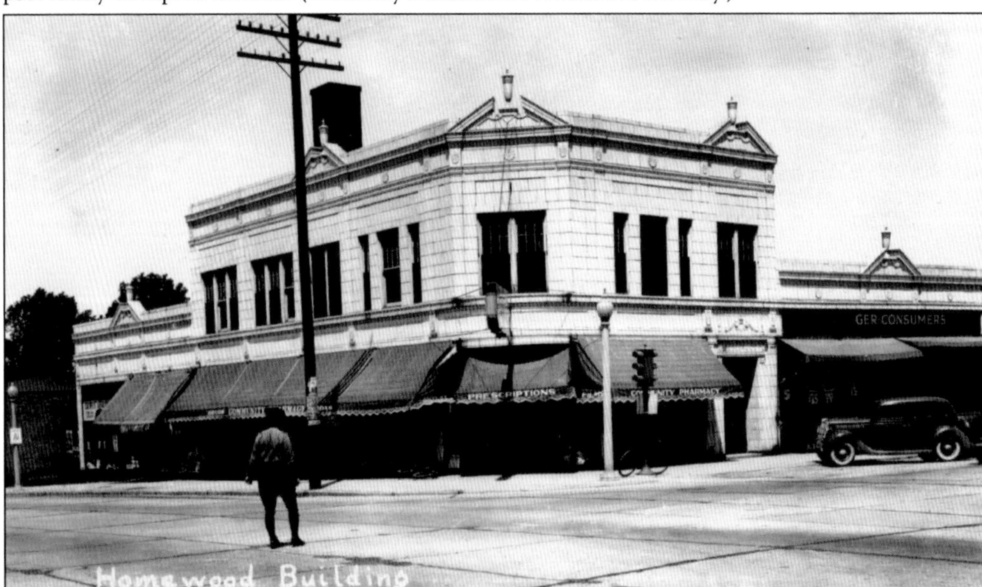

Increased traffic on the Dixie Highway in Homewood yielded further business improvements at the intersection with Ridge Road. The Homewood building, which is situated on the northeast corner, was completed in 1925, and for over 50 years, the Community Pharmacy occupied the first floor. The building has a beautiful white terra-cotta finish, and a travel agency is now the main tenant. (Courtesy Homewood Historical Society.)

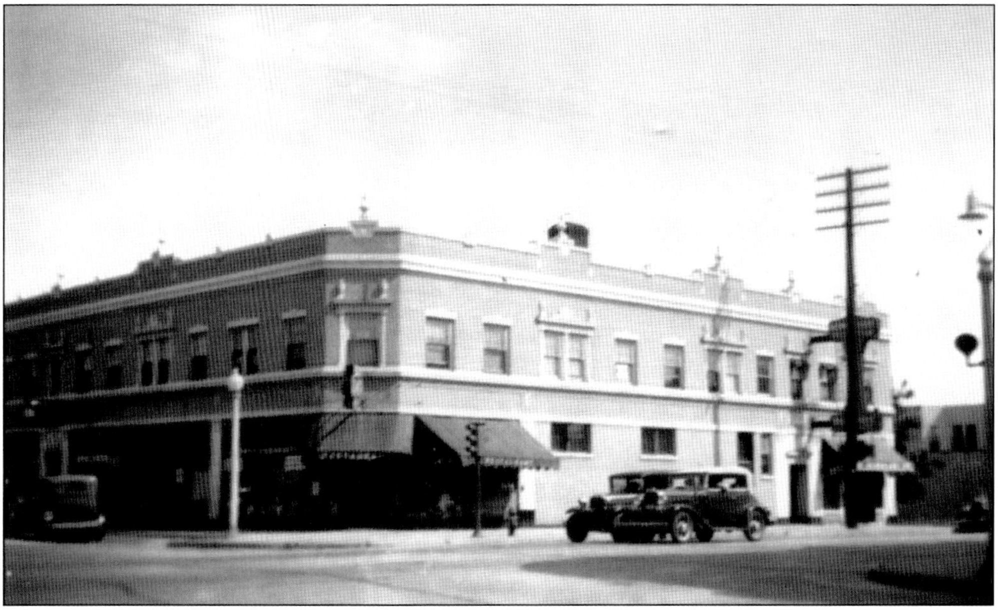

The Dixie-Ridge building sits on the southeast corner of its namesake intersection in Homewood. The building was formally opened on October 12, 1929, just weeks before the stock market crash. Despite the inauspicious beginnings, the structure is still a Homewood landmark, and after 80 years, a barbershop and restaurant continue to be main tenants. (Courtesy Homewood Historical Society.)

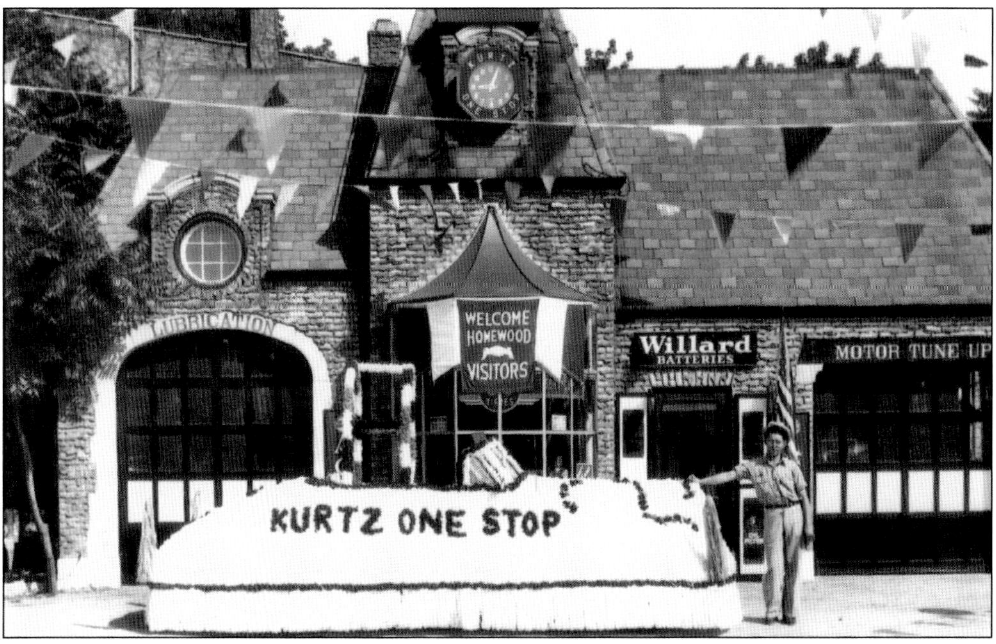

Jack Kurtz stands beside a float in front of his Sinclair One Stop service station. The quaint cottage-style station was located on the northwest corner of Ridge Road and the Dixie Highway in Homewood. A new station was built on this site in the 1950s and was owned by Kurtz until 1979. A succession of owners has operated the station since then selling gas under a variety of brands. (Courtesy Homewood Historical Society.)

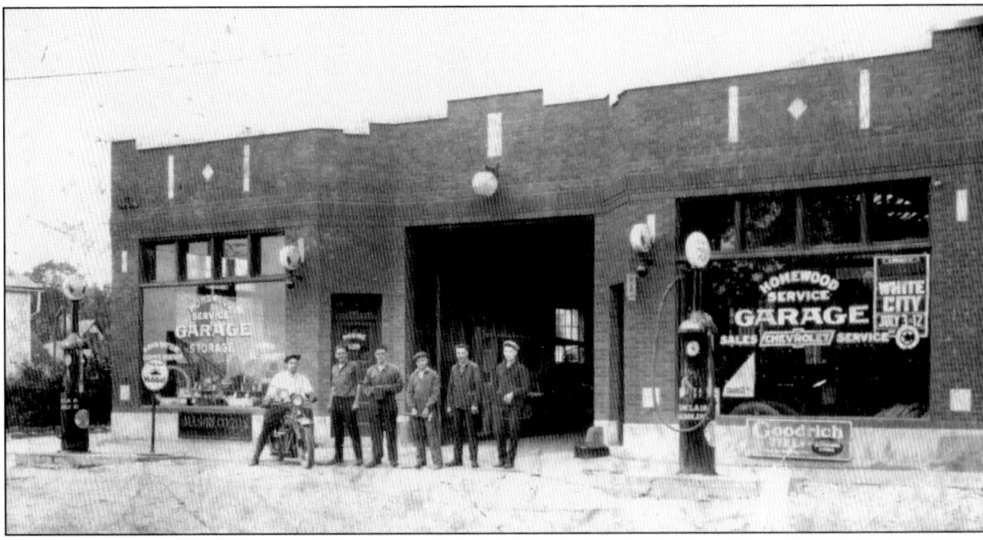

Bert Heuer Sr. (right) operated his Chicago Motor Club–approved service garage built in 1924 on the Dixie Highway just south of Ridge Road in Homewood. The garage sold Pennzoil products, Exide batteries, and Standard Red Crown gasoline dispensed from a convenient curbside pump. (Courtesy Homewood Historical Society.)

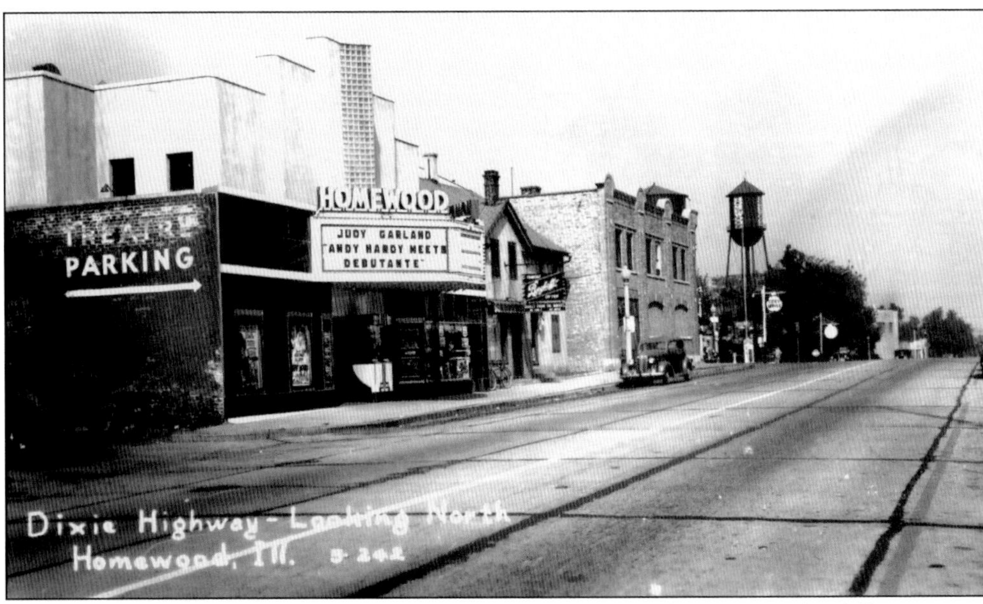

A victim of the Depression, the Heuer Service Garage closed in 1936. The following year, the garage was extensively renovated and converted into the 659-seat Homewood Theatre. The theater had all the latest amenities for its patrons and was one of the first buildings in town to be air-conditioned. The building was torn down in 1992. (Courtesy Homewood Historical Society.)

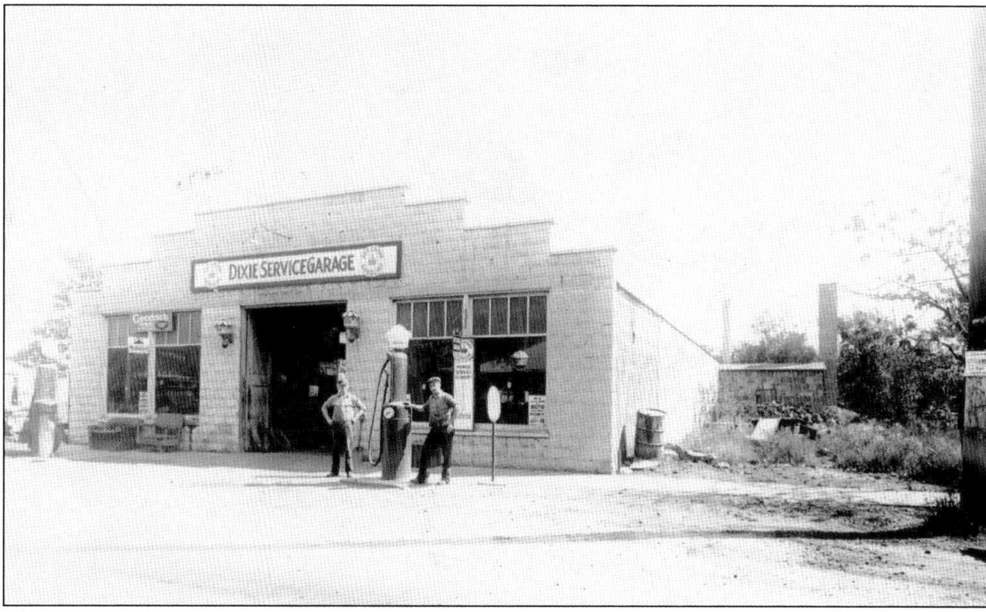

The Schmidt brothers operated the Dixie Service Garage at 18301 Dixie Highway in Homewood. The garage was later sold to Irvin Ruge, who operated the business for many years. Van Drunen Ford relocated its dealership to the building in 1949 and significantly enlarged the structure. The building burned down on Halloween night of 2004 and ironically was home to Chuck's House of Magic at the time. (Courtesy Homewood Historical Society.)

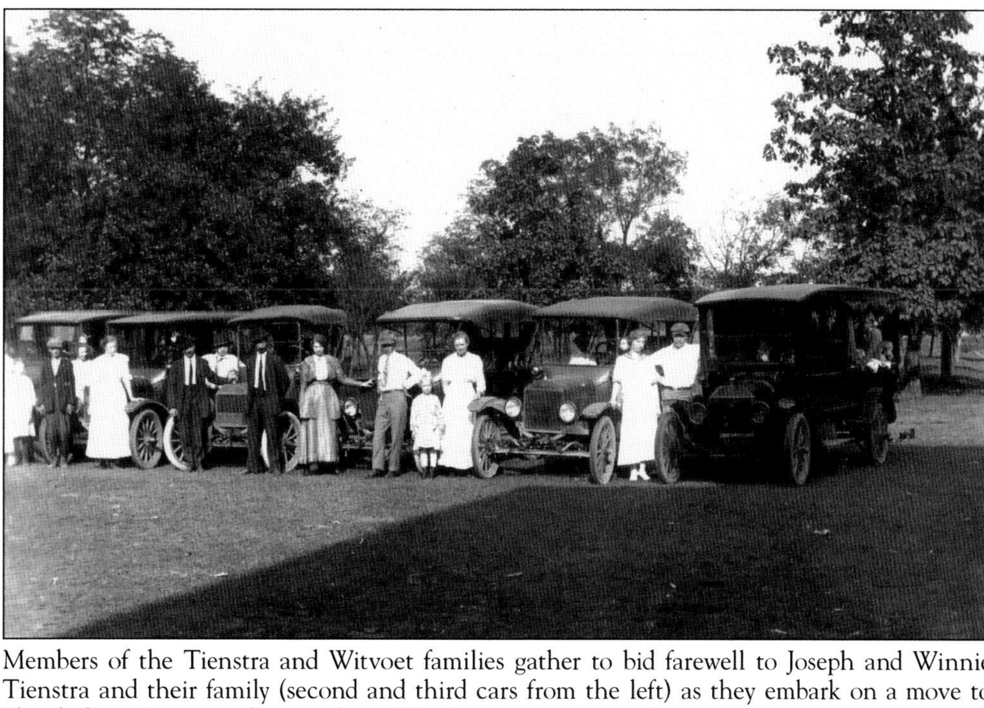

Members of the Tienstra and Witvoet families gather to bid farewell to Joseph and Winnie Tienstra and their family (second and third cars from the left) as they embark on a move to Florida from Homewood in October 1920. The Tienstra family followed the Dixie Highway on its trip and lived in Florida for four years before moving back to Homewood. Garrett Tienstra, who celebrated his 101st birthday in 2009, remembers the trip vividly. (Courtesy Garrett Tienstra.)

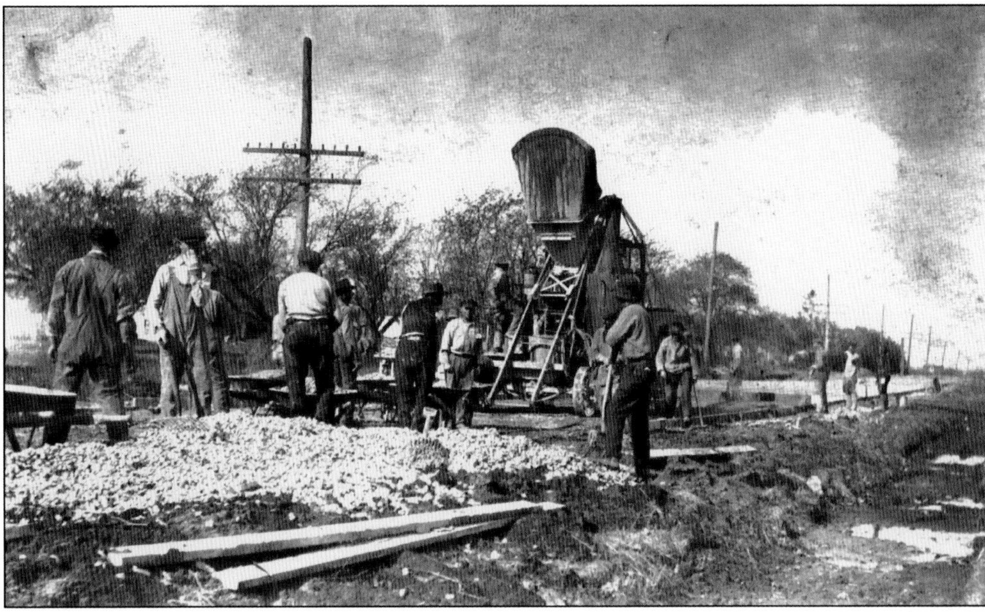

Paving crews from the firm of Pronger and Fletcher of Blue Island complete concrete work on the Dixie Highway in Flossmoor just south of the Homewood border in this view from the summer of 1917. The Hecht farmstead is in the background, and the old home still stands at 18916 Dixie Highway. (Courtesy Chuck Hecht.)

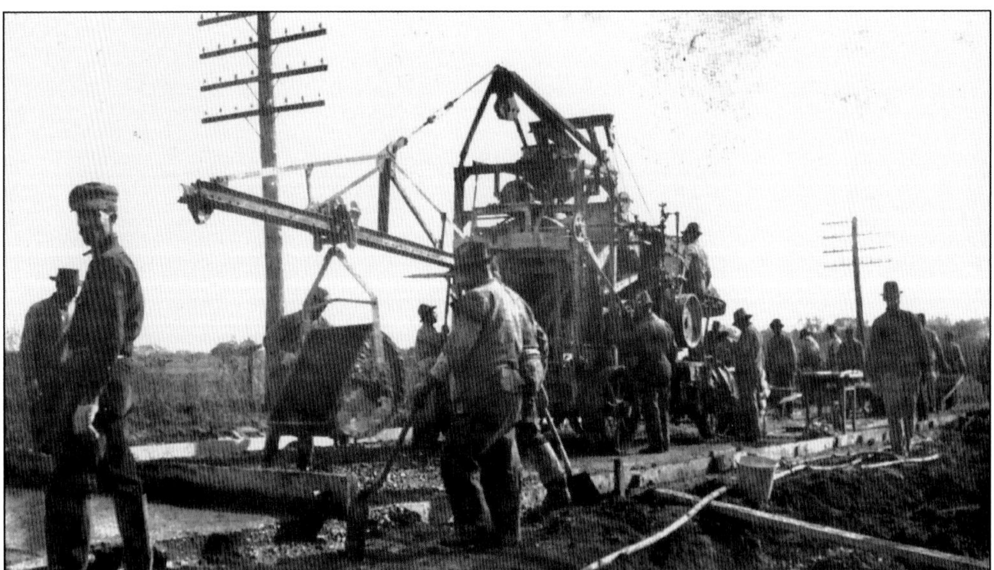

Another load of concrete is being dumped and ready for finishing in this photograph of Pronger and Fletcher paving crews working on the Dixie Highway in Flossmoor in 1917. Although the Hecht farm has been developed into a suburban subdivision, the Idlewild and Flossmoor Country Clubs still flank the highway as it wends its way through Flossmoor. (Courtesy Chuck Hecht.)

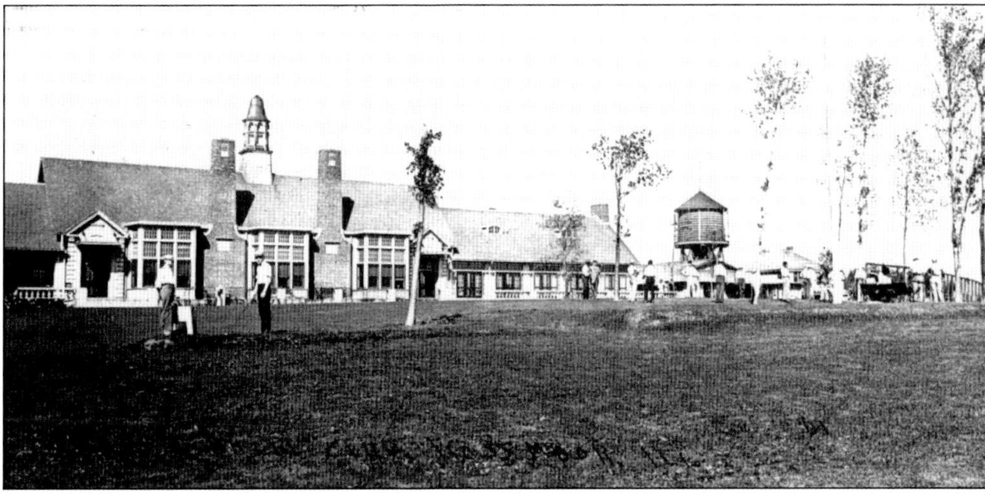

Flossmoor Country Club lies just west of the Dixie Highway, south of Flossmoor Road. The first two clubhouses built on the golf course were destroyed by fire in 1907 and 1914. The present clubhouse was completed in 1916. The club, established in 1899 as the Homewood Country Club, is one of the oldest in Chicagoland. (Author's collection.)

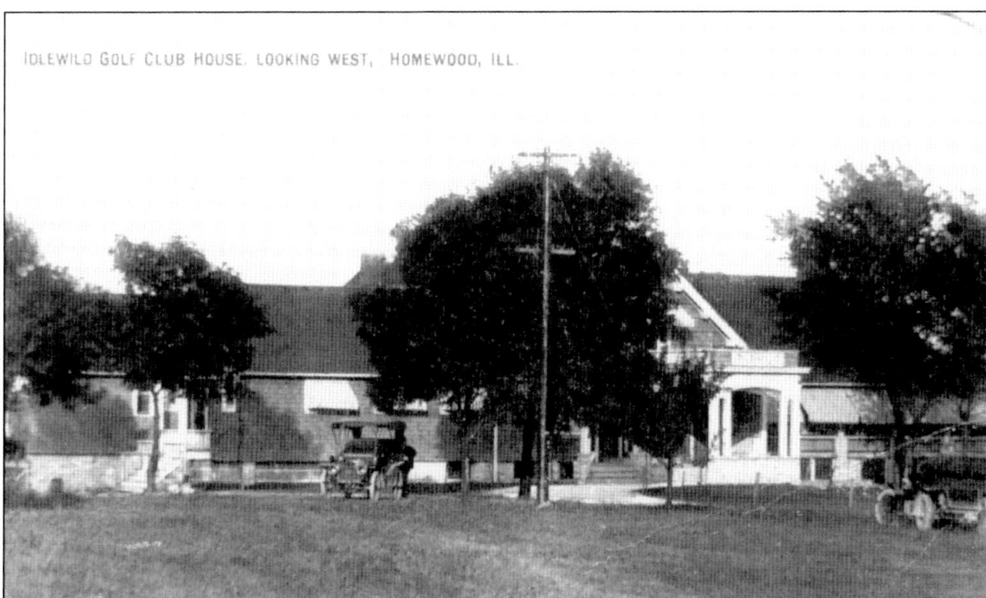

Idlewild Country Club, on the east side of the Dixie Highway at Flossmoor Road in Flossmoor, was established in 1908. The club occupies 156 acres of land originally homesteaded by area pioneer Benjamin Butterfield. A creek that bears his name meanders through the course, and a bridge, built in 1930, fords the creek on the Dixie Highway. (Author's collection.)

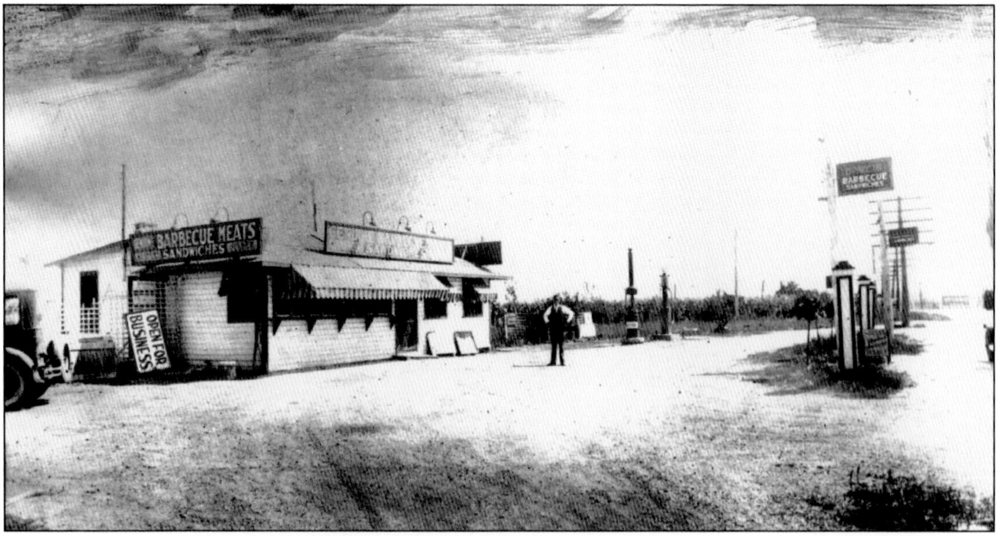

John Milekich stands in front of his barbecue stand in Chicago Heights on the east side of the Dixie Highway, just north of Joe Orr Road. Milekich sold sandwiches, ice cream, cold drinks, cigars, cigarettes, and gas. The building was later torn down, and a new structure built on the site housed Ada's Restaurant for years. (Courtesy Chicago Heights Public Library.)

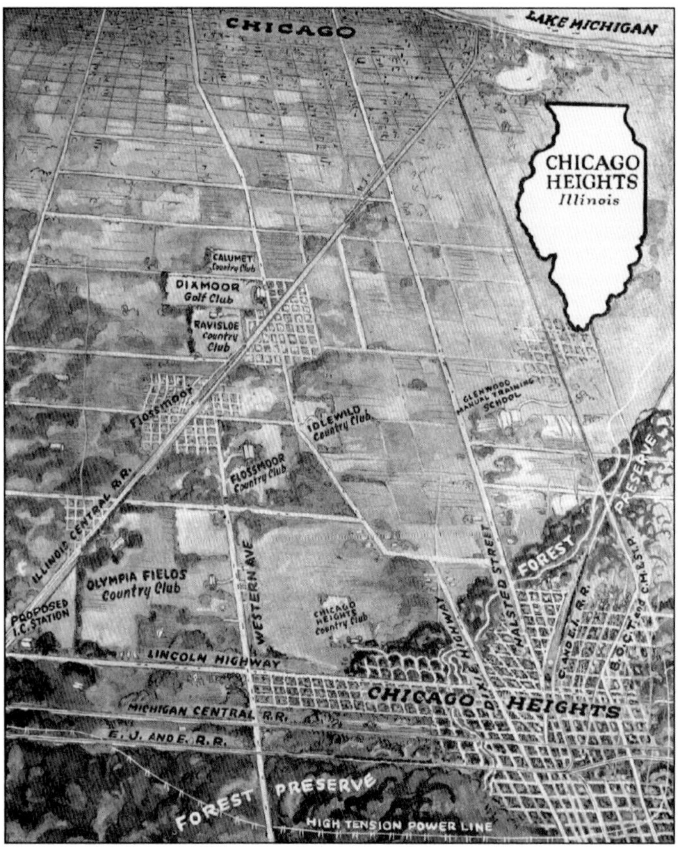

This 1925 map, published by the Chicago Heights Chamber of Commerce, shows the prominence of the city in the south suburban area. Of note, it illustrates the original route of the Dixie Highway through Chicago Heights. At this time, traveling south, the highway followed Joe Orr Road east to Chicago Road, then south through Chicago Heights. (Courtesy Dominic Candeloro.)

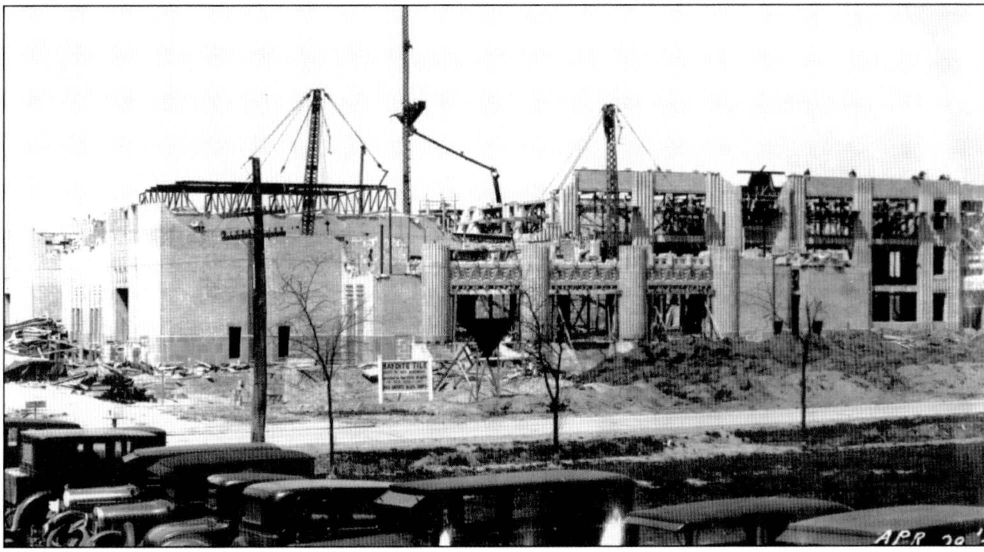

Construction on Bloom High School in Chicago Heights was well underway when this photograph was taken in April 1931. The art deco–style building was built in the depths of the Depression and officially opened in February 1934. The structure, a Dixie Highway landmark, is listed on the National Register of Historic Places. (Courtesy Bloom High School Archives.)

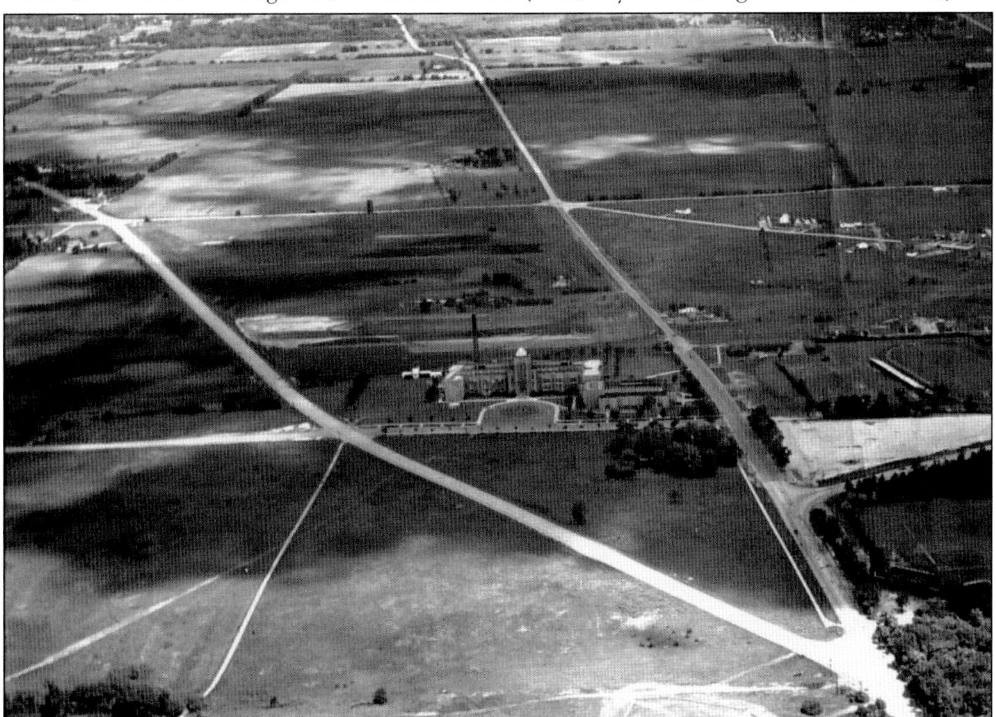

This aerial view of Chicago Heights shows Bloom High School (center), Chicago Road (right), and the Dixie Highway angling off to the left. The Dixie Highway cutoff eliminated the right-angle turns of the original alignment on Joe Orr Road and Chicago Road. The easement for the cutoff was recorded in December 1929; however, the roadway was not built until the mid-1930s. (Courtesy Chicago Heights Public Library.)

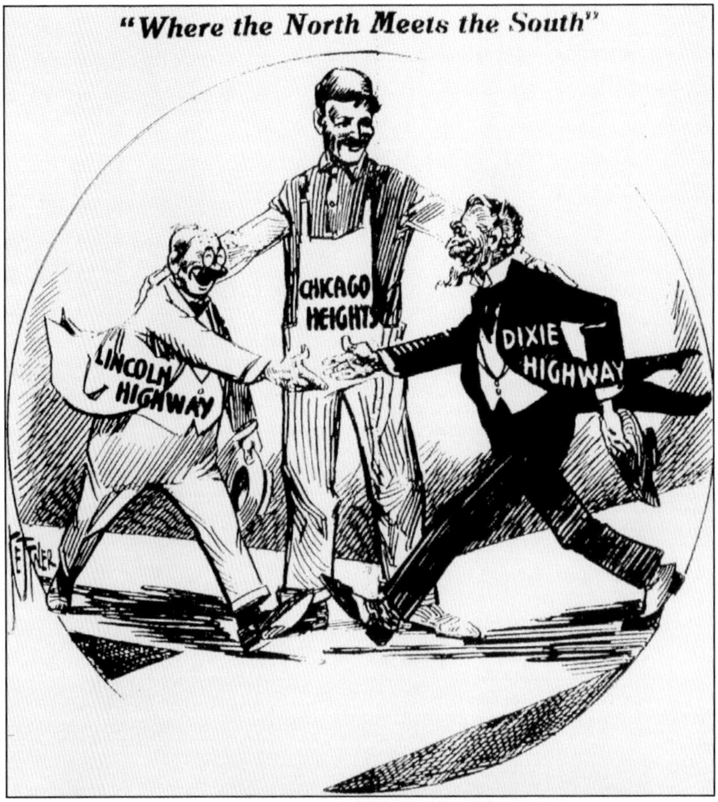

As this cartoon from 1915 indicates, Chicago Heights touted itself as "Where the North Meets the South." Today the city proudly proclaims itself as the "Crossroads of the Nation." No one would have argued in 1915 that Chicago Heights was at the crossroads of the nation's two most prominent roadways—the Lincoln and Dixie Highways. (Author's collection.)

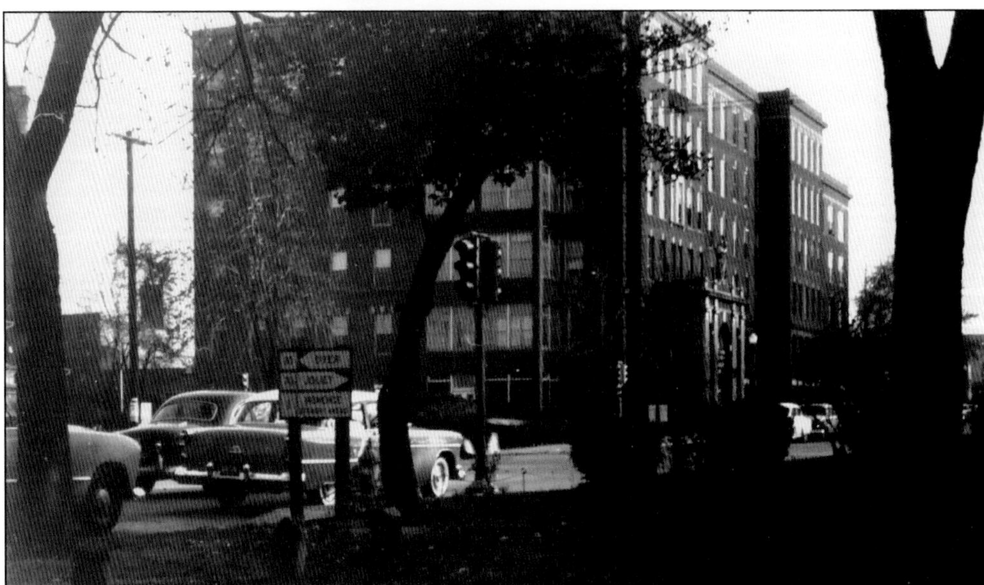

The original St. James Hospital building occupied land near the southeast corner of the Lincoln and Dixie Highways in Chicago Heights. The 50-bed hospital was dedicated on Thanksgiving Day 1911. The facility has been greatly enlarged over the years and has been operated under the direction of the Sisters of St. Francis of Lafayette, Indiana. (Courtesy Chicago Heights Public Library.)

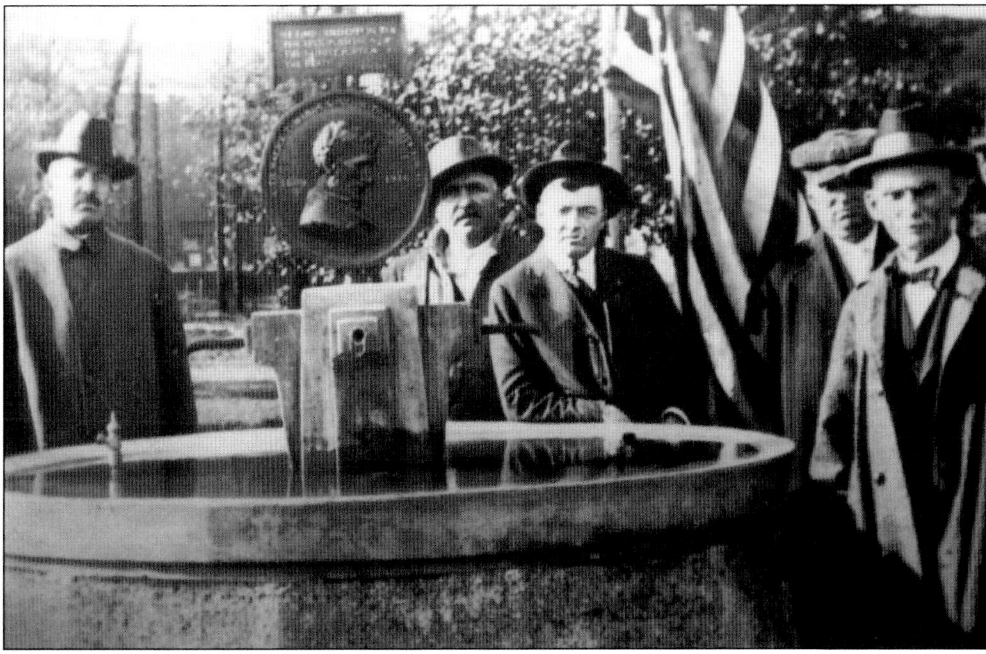

The Arche Fountain was a gift to Chicago Heights from the Arche Club of Chicago. The fountain, dedicated on October 13, 1916, is located on the southwest corner of the Lincoln and Dixie Highways and prominently features a replica of a Lincoln penny. (Courtesy Chicago Heights Public Library.)

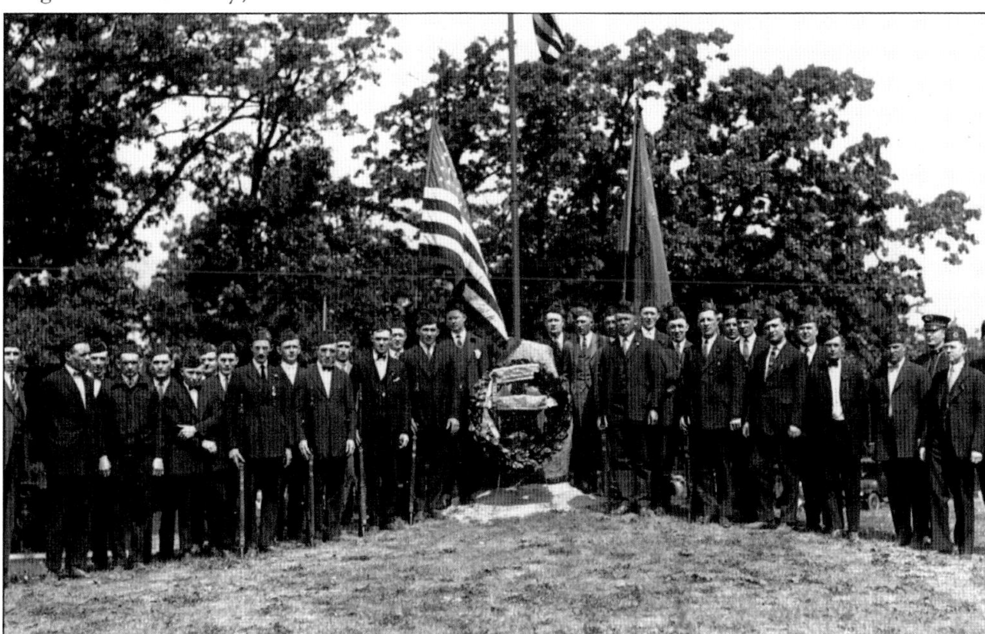

At the northwest corner of the Lincoln and Dixie Highways, the soldiers' monument was dedicated on May 20, 1922. The monument grounds have been enhanced since that time and serve as the gathering spot for Chicago Heights residents for Memorial Day, Veterans Day, and Christmas tree lighting ceremonies. (Courtesy Chicago Heights Public Library.)

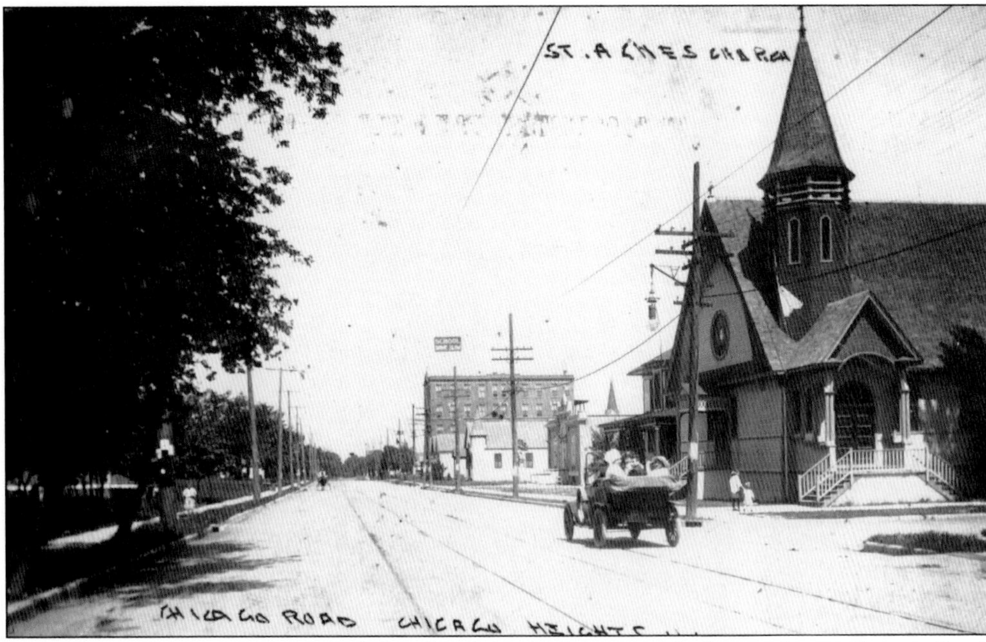

This view from late in the second decade of the 20th century looks north on Chicago Road, the Dixie Highway, from McEldowney Place in Chicago Heights. The original church, rectory, and school of St. Agnes Parish are pictured immediately from right to left in the photograph. St. James Hospital is located in the distance. Note the interurban tracks and the horse and buggy. (Courtesy George Albrecht.)

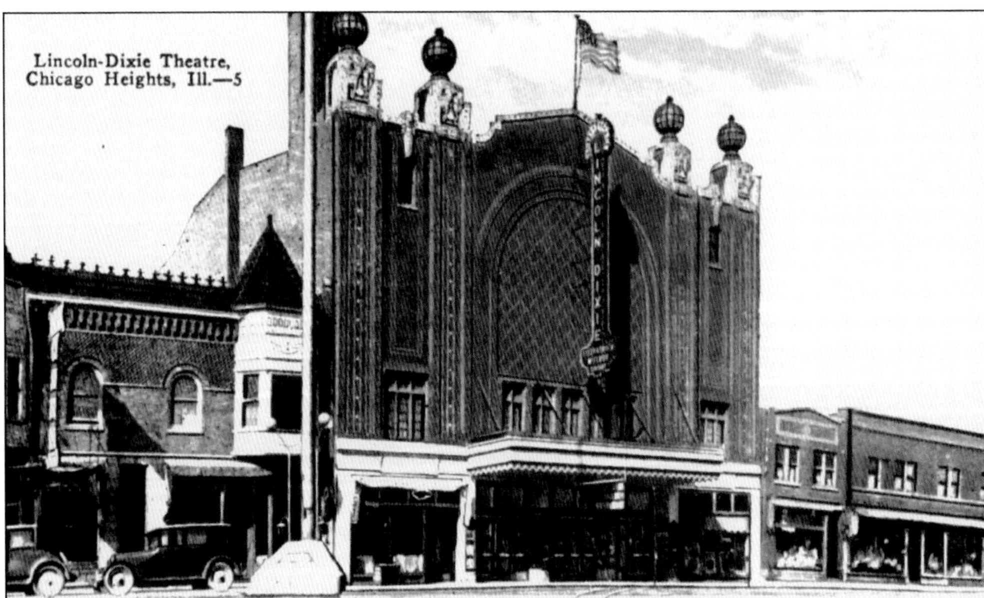

Built at a cost of $300,000, the Lincoln-Dixie Theatre opened in Chicago Heights on June 23, 1921, and was located at 1658 Chicago Road, the Dixie Highway. The theater could seat 1,800 and was considered one of the finest in the Chicago area. The theater closed after a half century of operation and was demolished in 1972. (Courtesy Chicago Heights Public Library.)

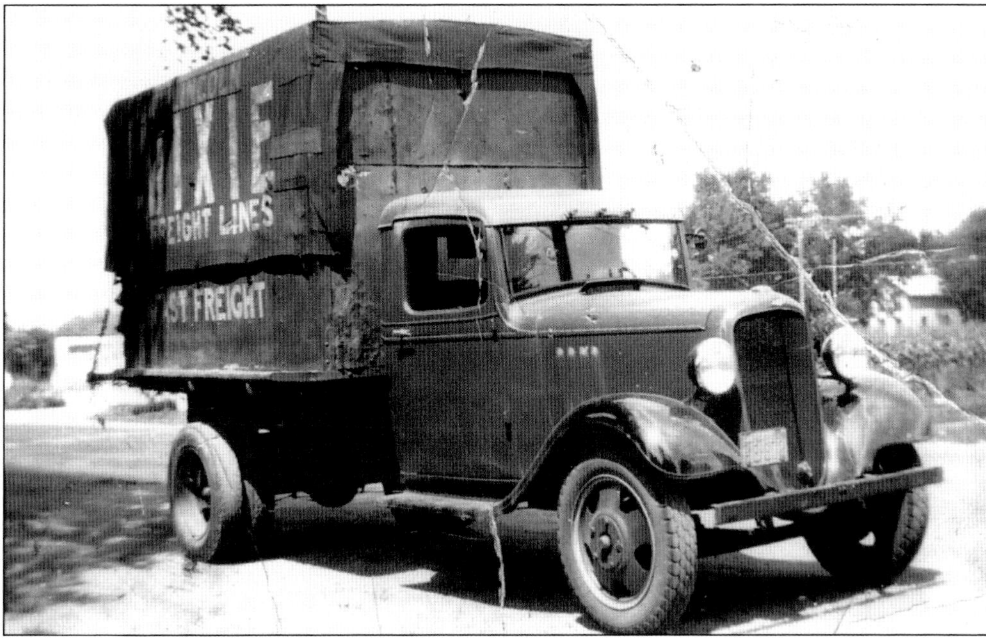

Andrew Leoni started his trucking company in 1930 with headquarters in Chicago Heights. He aptly named the firm Lincoln-Dixie Freight Lines. The company changed its name to Leoni Motor Express in 1935 and still operates out of offices located on Ashland Avenue. (Courtesy Mike Leoni.)

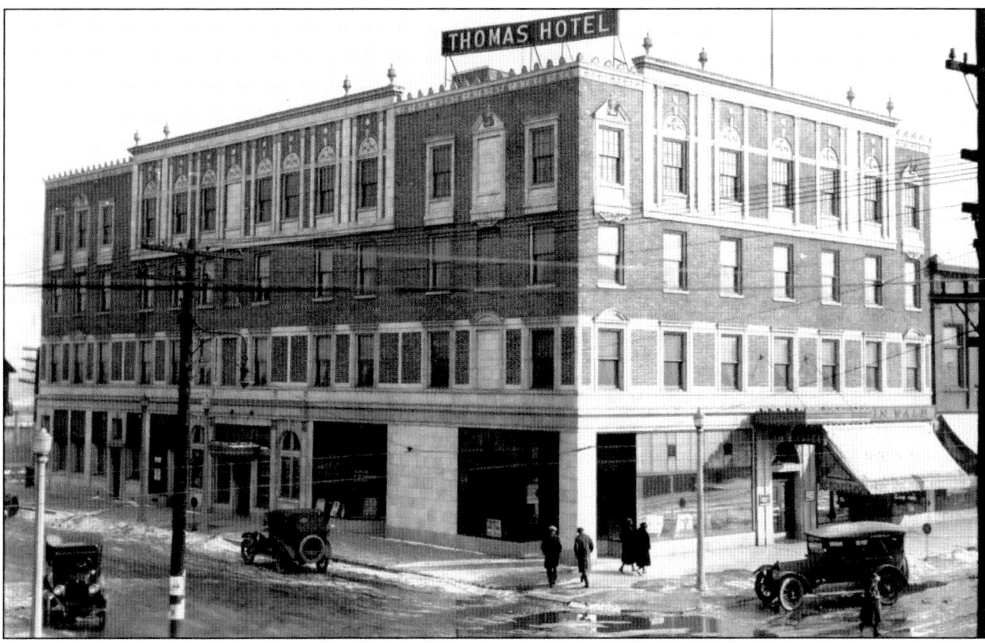

Although not on the Dixie Highway, the Thomas Hotel in Chicago Heights was located just east of the highway on the southeast corner of Seventeenth and Halsted Streets. The hotel, built in 1923, offered travelers luxurious accommodations with up-to-date amenities. It later became known as the Albert Hotel and was demolished after a fire in 1979. (Courtesy Dominic Candeloro.)

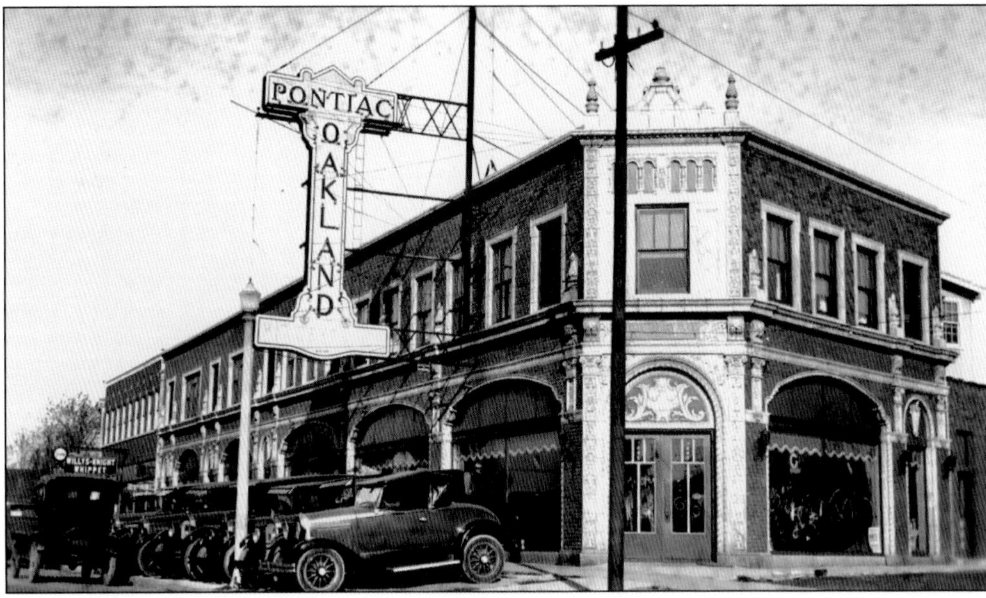

H. C. Wehrmann opened his Pontiac and Oakland dealership in the Flat Iron Building in Chicago Heights on Emerald Avenue in 1925. Wehrmann was a prominent automobile dealer and also had dealerships at this time in Goodenow and Grant Park, two other Dixie Highway communities. The Flat Iron Building was listed on the National Register of Historic Places in 2003. It was demolished in 2009. (Courtesy Dominic Candeloro.)

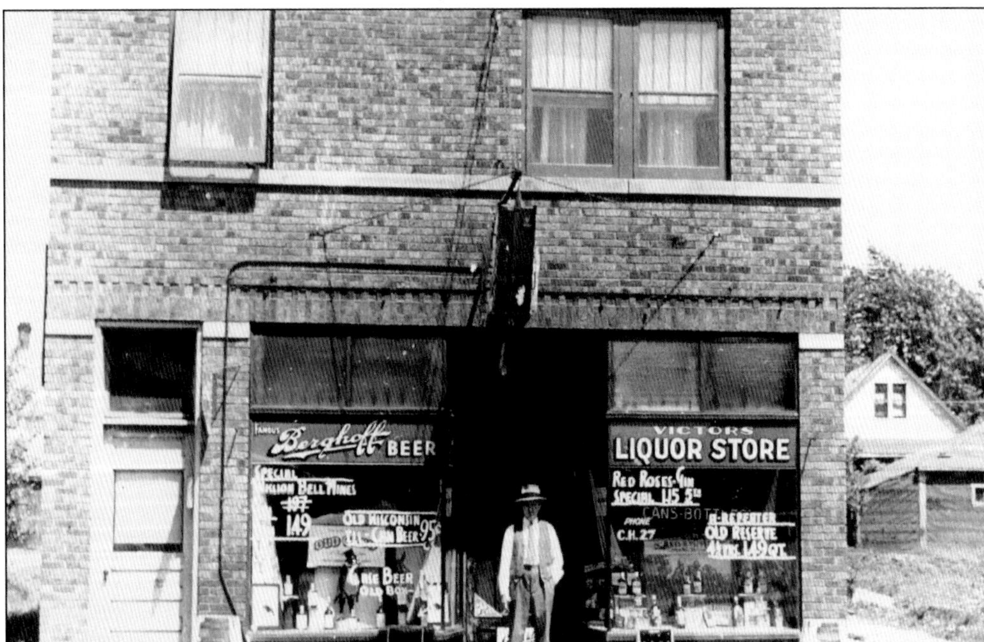

Victor Martello stands in front of Victor's Liquor Store located in South Chicago Heights at 2619 Chicago Road, the Dixie Highway, in this view from the late 1930s. In addition to Berghoff beer, Martello advertises Red Roses gin for $1.45 a fifth and Old Reserve aged whiskey for $1.49 a quart. The building still stands on the east side of Chicago Road. (Courtesy South Chicago Heights Historical Committee.)

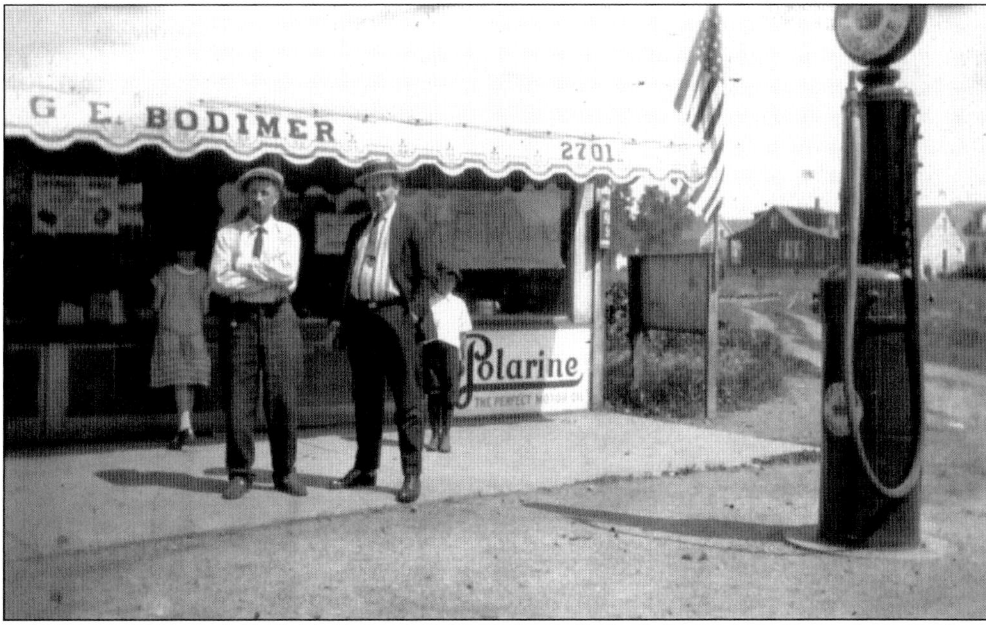

Customers stand in front of the G. E. Bodimer store at 2701 Chicago Road in South Chicago Heights. Bodimer sold Standard Red Crown gasoline along with Standard's Polarine motor oil curbside. Inside, a hand-lettered sign in the window advertises fireworks, while on the opposite side of the store, another sign touts Lifebuoy and Rinso soaps as among the many products Bodimer peddled. (Courtesy South Chicago Heights Historical Committee.)

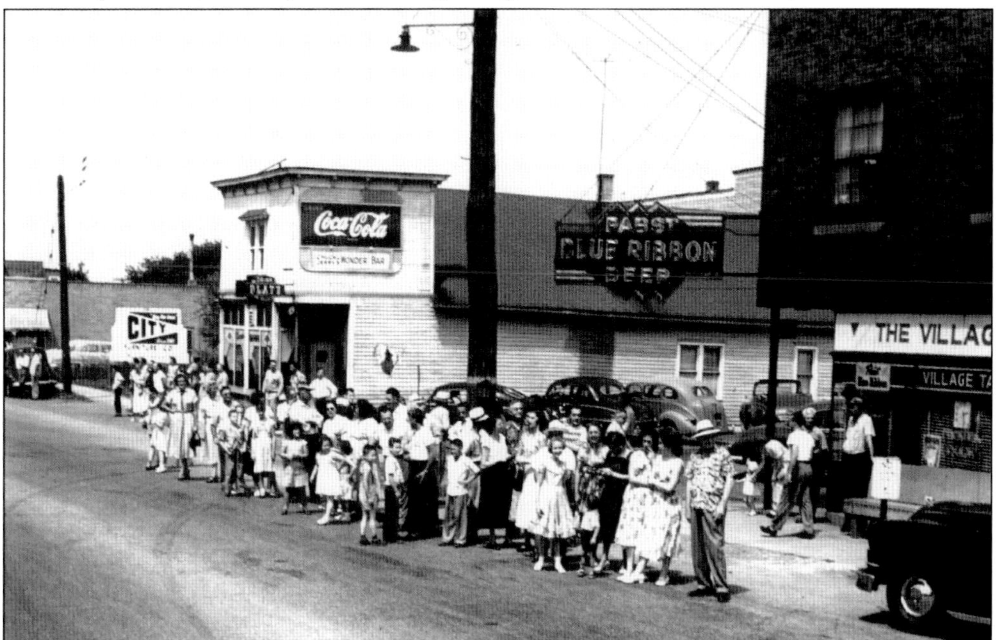

Residents await the start of the South Chicago Heights 1948 homecoming parade. The group is standing on the east side of Chicago Road at Twenty-seventh Street. At this time, Chuck and Jerry's Wonder Bar occupies the northeast corner, while the Village Tap occupies the southeast corner of the intersection. (Courtesy South Chicago Heights Historical Committee.)

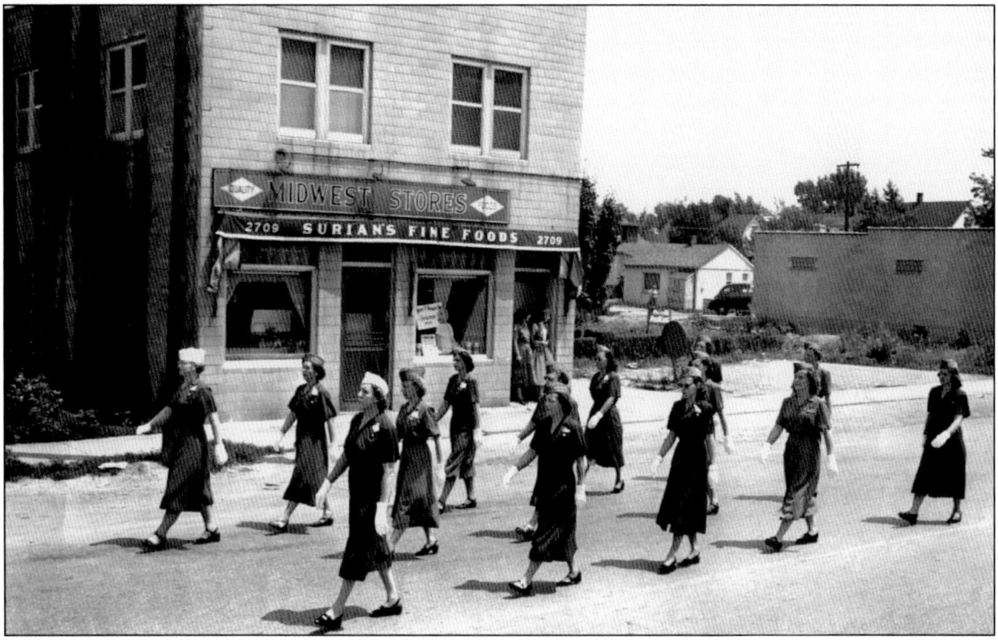

The women's auxiliary for a veterans organization marches north on Chicago Road during the 1948 homecoming parade in South Chicago Heights. The group is seen in front of Surian's Fine Foods, a Midwest Stores grocer located at 2709 Chicago Road. The building, of a distinctive cream-colored enamel brick, still stands. (Courtesy South Chicago Heights Historical Committee.)

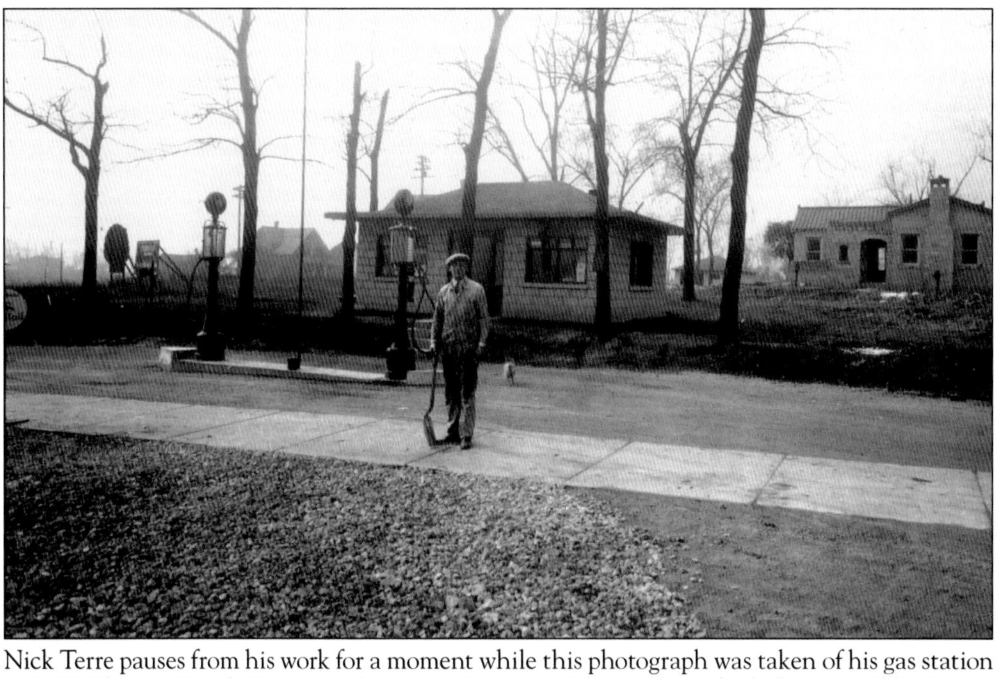

Nick Terre pauses from his work for a moment while this photograph was taken of his gas station at 2906 Chicago Road. Terre was the original owner of the station, which dates to 1927. He was also the village president of South Chicago Heights from 1935 to 1941. The station still stands and was last occupied by a hair salon. (Courtesy South Chicago Heights Historical Committee.)

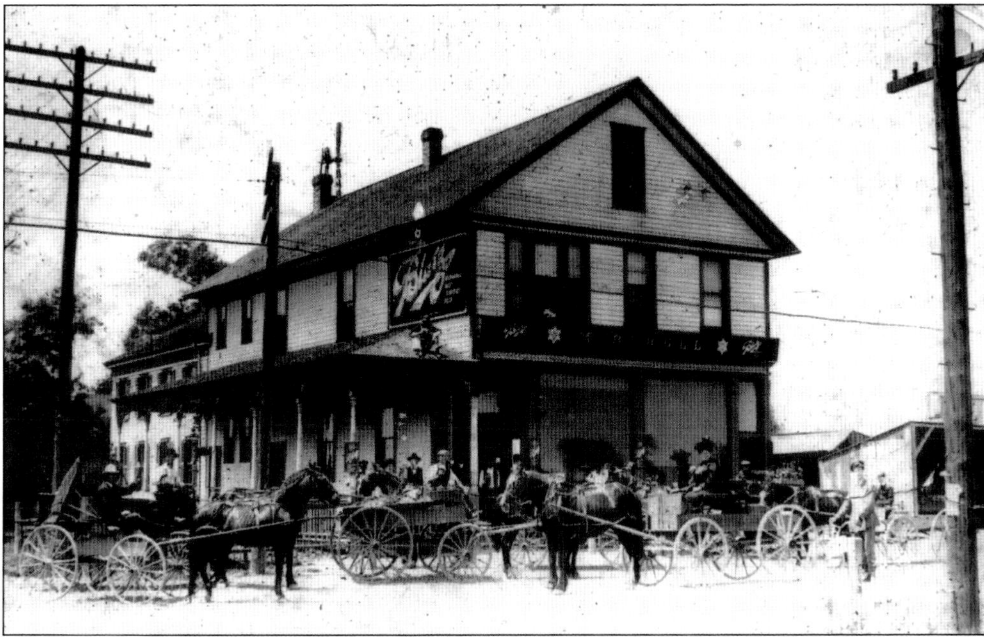

Adam Brown was the first settler in what was to become Bloom Township, arriving in 1833. He eventually owned land on all corners of the Sauk and Hubbard's Trails. The intersection became known as Brown's Corner. By the beginning of the 20th century, Michael Burgel was operating a tavern and hotel out of the old Brown home in South Chicago Heights. (Courtesy South Chicago Heights Historical Committee.)

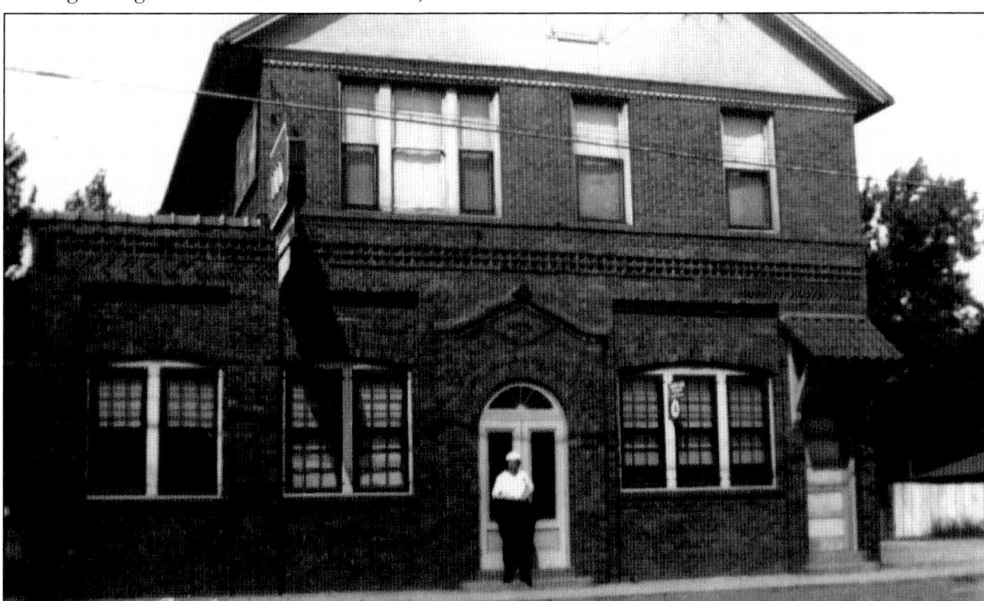

Mary Burgel-Barman took over her father's tavern and restaurant after his death. The Burgel House was a popular South Chicago Heights dining establishment and was updated and remodeled over the years. By the 1940s, the exterior had been veneered with brick and bore little resemblance to the old Brown home. The building was torn down in 1968 for a gas station. A Walgreens drugstore now occupies the site. (Courtesy South Chicago Heights Historical Committee.)

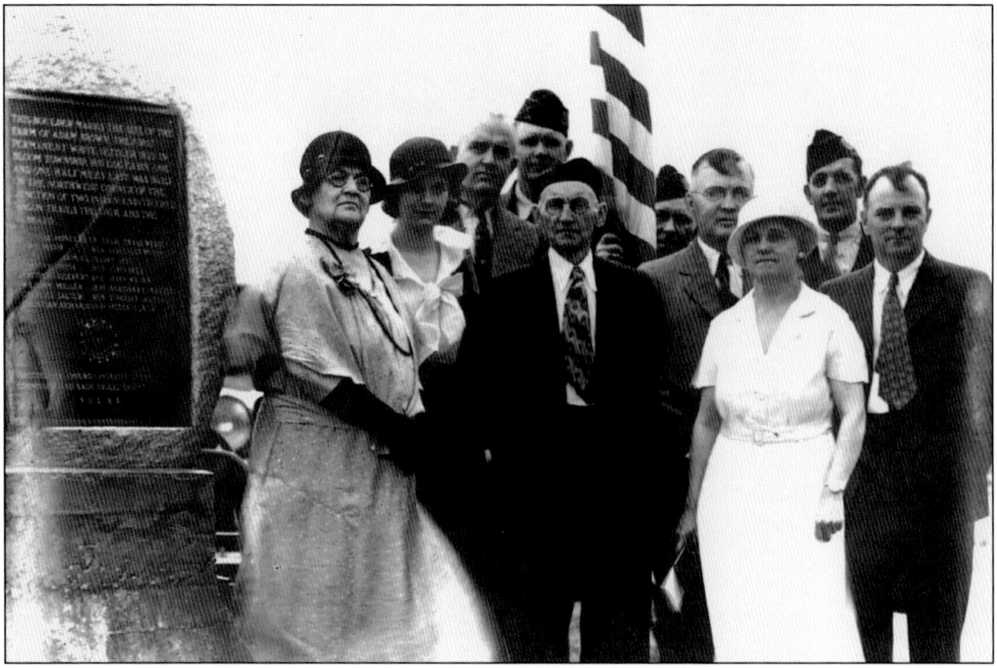

Adam Brown Jr. (center), the only surviving son of Bloom Township's first permanent settler, Adam Brown, attended this 1933 South Chicago Heights ceremony dedicating a boulder to mark the site of his father's farm. The plaque and boulder were located at the intersection of the Sauk Trail and Chicago Road, the Dixie Highway. (Courtesy of South Chicago Heights Historical Committee.)

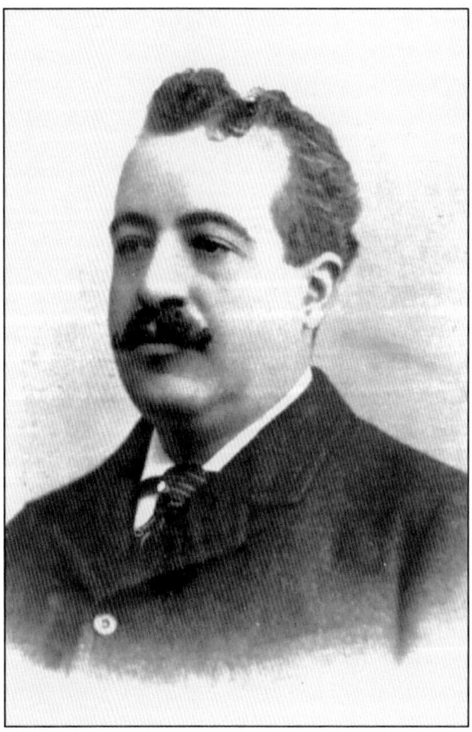

John Steger was president of the Steger and Sons Piano Manufacturing Company and the namesake of the village of Steger. The citizens of the village bestowed this honor on him after he paid all the costs associated with the village's incorporation and first election. Steger was born in Germany in 1854 and came to the United States at the age of 17 and died in 1916. (Courtesy Steger Historical Society.)

Steger and Sons not only produced pianos; by 1922, it was also producing phonographs in fine wood cabinetry. The company's headquarters were in the Steger Building at the corner of Jackson Boulevard and Wabash Avenue in downtown Chicago, and the company's factories were on Chicago Road, the Dixie Highway, in Steger. The company's advertising touts Steger and Sons as being "where the 'Lincoln' and 'Dixie' Highways meet." (Courtesy Steger Historical Society.)

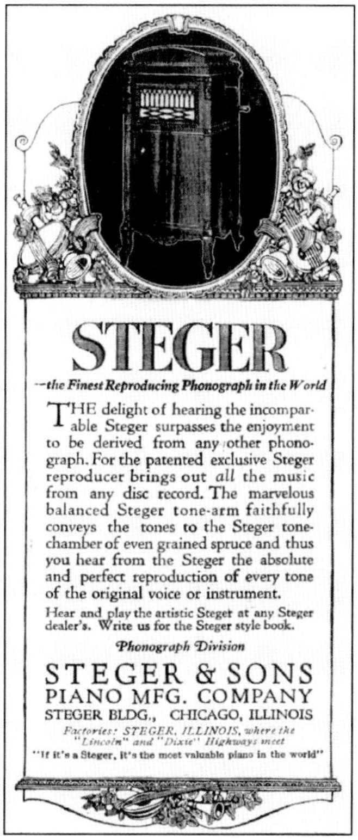

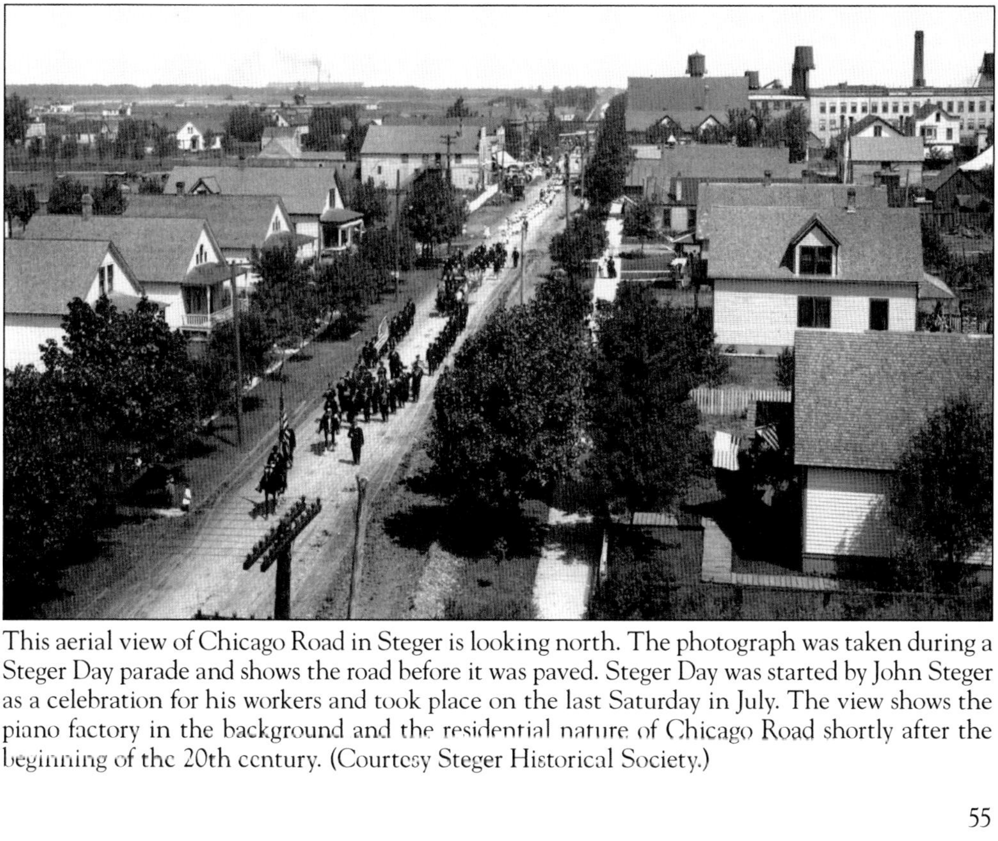

This aerial view of Chicago Road in Steger is looking north. The photograph was taken during a Steger Day parade and shows the road before it was paved. Steger Day was started by John Steger as a celebration for his workers and took place on the last Saturday in July. The view shows the piano factory in the background and the residential nature of Chicago Road shortly after the beginning of the 20th century. (Courtesy Steger Historical Society.)

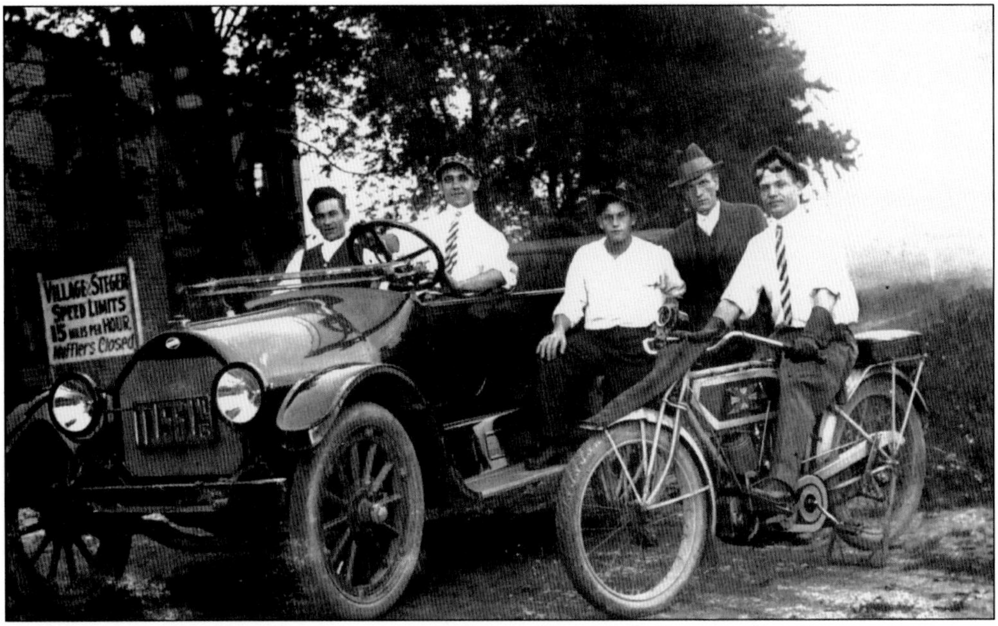

The latest in motor equipment is pictured in this 1915 photograph taken in Steger on Thirty-fourth Street east of the Dixie Highway. The Excelsior Auto Cycle was a product of a company that Ignaz Schwinn of bicycle fame purchased and developed into the third-largest motorcycle manufacturer in the United States. The picture shows Steger's welcome sign, which advises of the village's 15 miles per hour speed limit and warns motorists their mufflers should be "closed" to keep noise at a minimum. (Courtesy Steger Historical Society.)

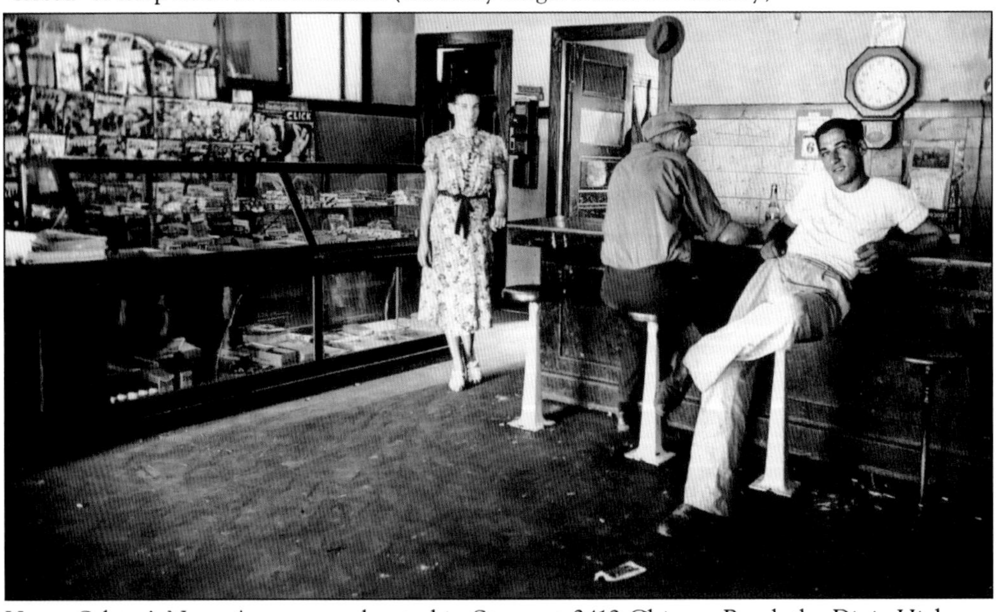

Henry Scheer's News Agency was located in Steger at 3412 Chicago Road, the Dixie Highway. In addition to newspapers and magazines, cigarettes, cigars, tobacco, and candies were sold here along with ice cream, sodas, coffee, and sandwiches that were served at the counter. Clerk Mildred Kuetemeyer is pictured with customer Buddy Maciariello facing the camera. Note the flypaper above the counter. (Courtesy Steger Historical Society.)

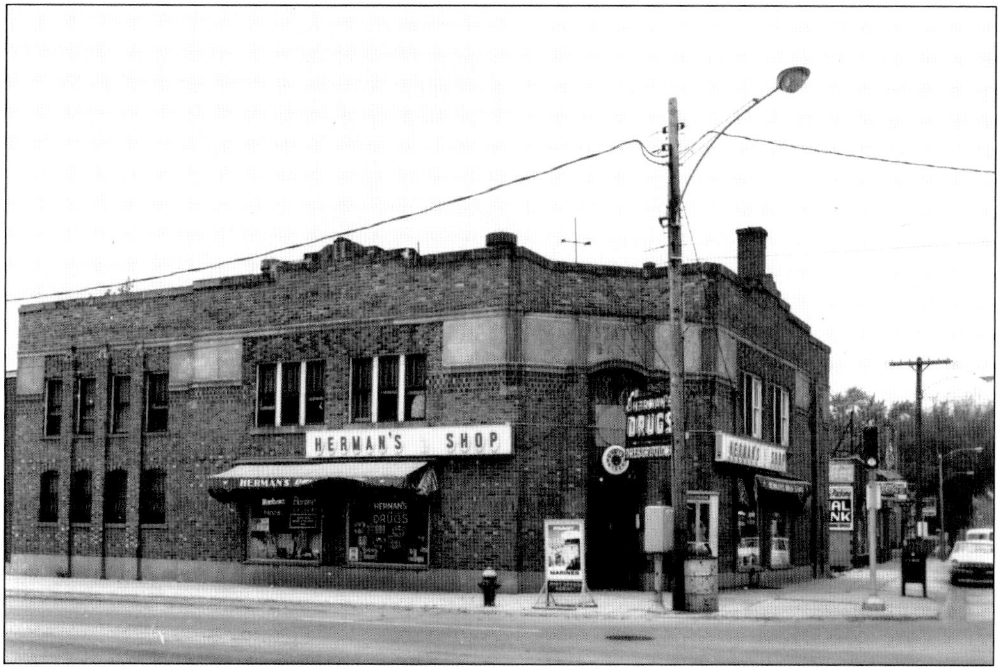

The First State Bank of Steger building was constructed in 1921 at a cost of $75,000 on the northwest corner of Chicago Road and Thirty-fourth Street. After the stock market crash in 1929, the bank went into receivership in 1932 and never reopened. A variety of businesses have called the bank building home since, including the Steger Sweet Shop and Herman's Drugs. (Courtesy Steger Historical Society.)

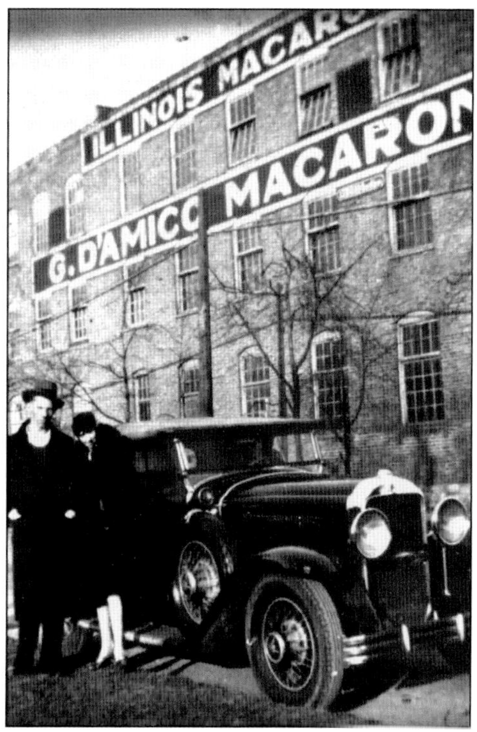

Tom Preble and friend are pictured near the D'Amico Macaroni plant at the southeast corner of Thirty-fifth Street and Chicago Road in Steger. In 1930, D'Amico purchased a Steger piano building that had been closed since 1922. D'Amico made pasta in the plant under the Mama Mia brand. In 1970, a new building was built one-half block south on Chicago Road. The old factory was destroyed by fire in 1974. A Shell gasoline station is now located on the site. (Courtesy Steger Historical Society.)

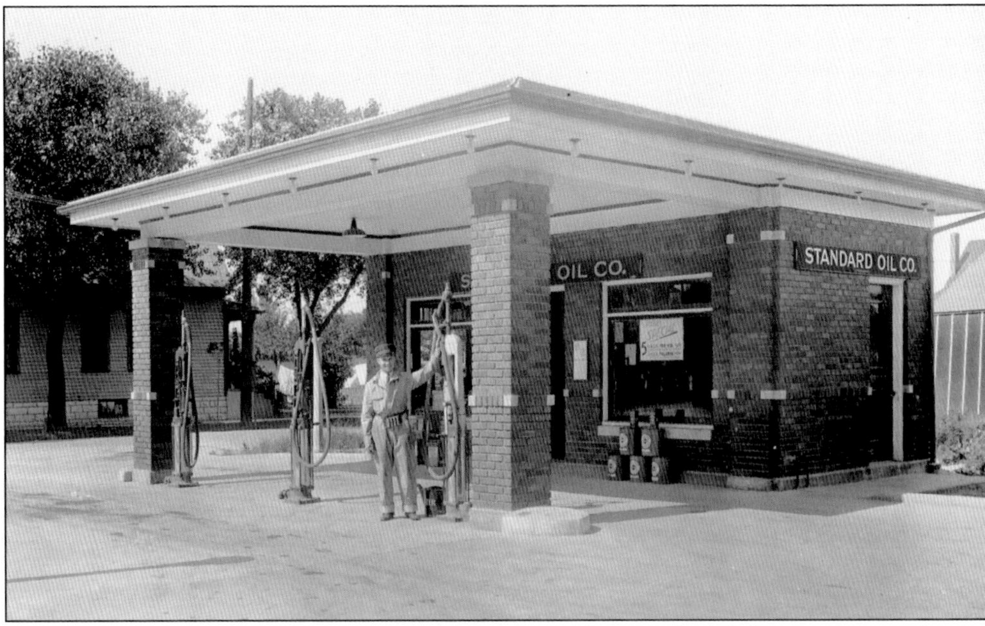

Harry Wichmann is shown standing near the pumps at his Standard Oil service station located in Steger at the northwest corner of Thirty-fourth Place and Chicago Road. The station was built in 1929 and has had a number of owners and uses over the years. The building still stands and remarkably retains much of its original look. A hair salon and now a used car dealer have been recent occupants. (Courtesy Steger Historical Society.)

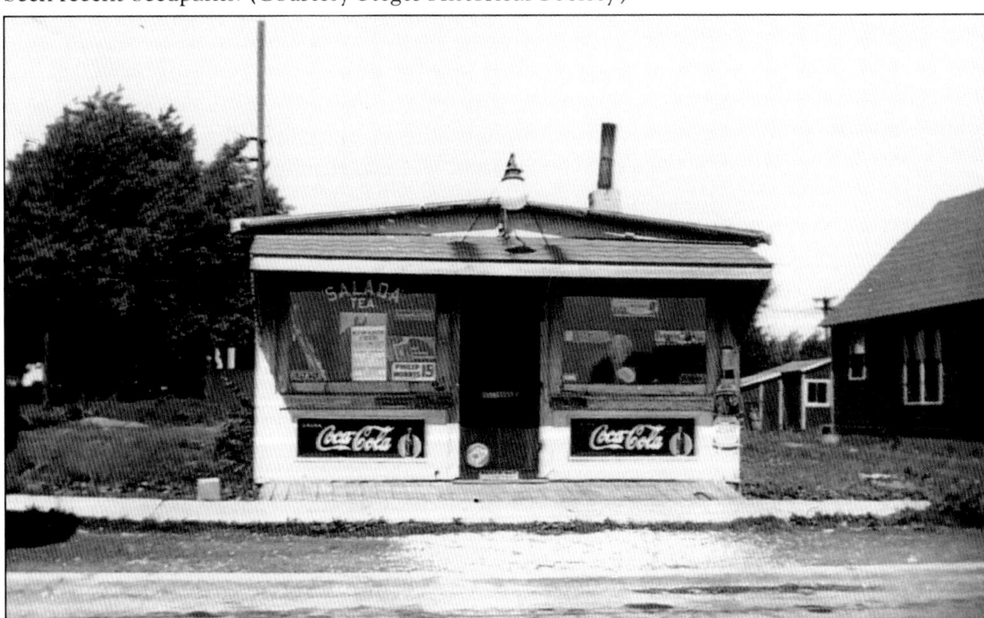

Alice Giebel operated this neighborhood store in Steger at Thirty-seventh Street and Chicago Road, the Dixie Highway, during the 1930s and early 1940s. Small stores like this, festooned with advertising for products like Coca-Cola, Pepsi-Cola, Salada tea, Mail Pouch tobacco, and Philip Morris cigarettes, were commonplace along the highway during this time frame. (Courtesy William and Miriam Giebel.)

Two

WILL COUNTY

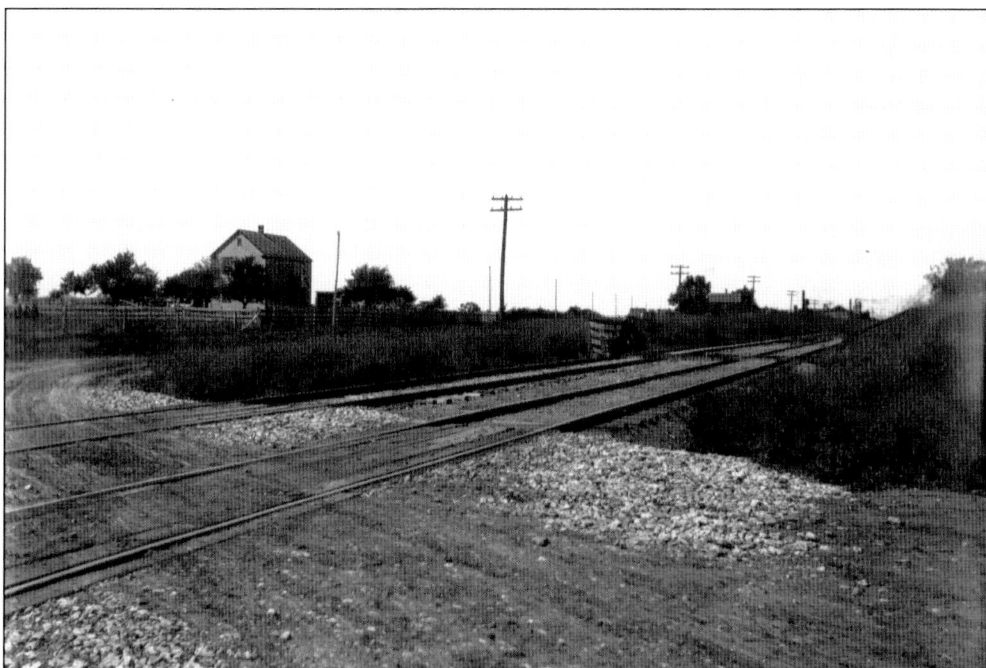

Until 1922, Illinois Route 1, the Dixie Highway, crossed the Chicago and Eastern Illinois Railroad (C&EI) tracks at grade near Richton Road, in Crete south of the Will County line. The unguarded crossing was very dangerous and was the scene of numerous accidents, many of which involved fatalities. Construction of the current railroad viaduct began in August 1922. This early view is looking north. (Courtesy Carol Triebold.)

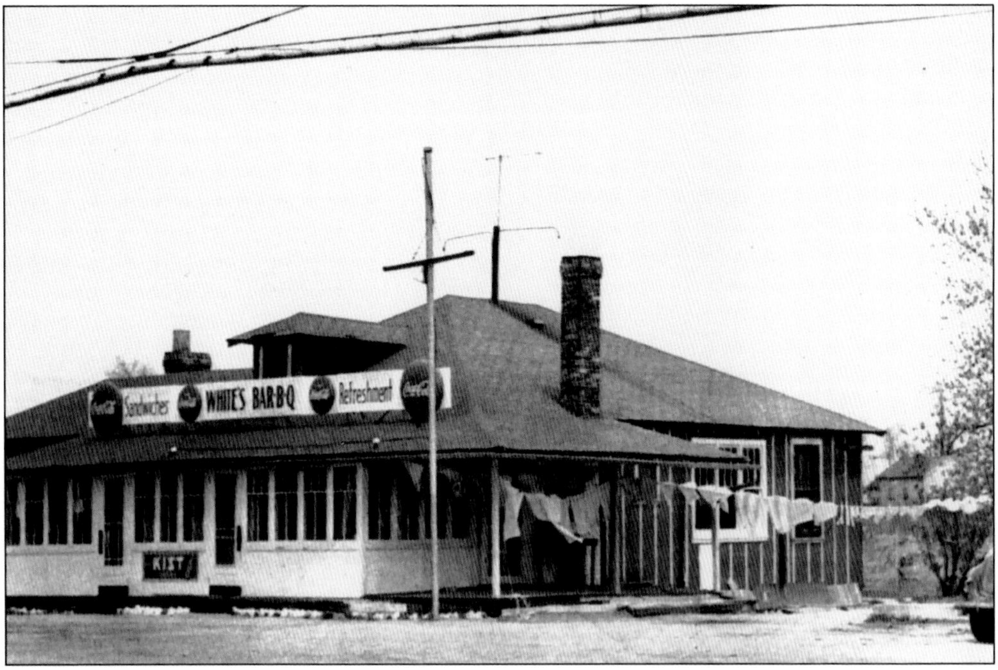

Joe White's barbecue stand was one of the early restaurants catering to motorists as they traveled through Crete. As the wash hanging in the photograph attests, the owner also lived on the premises. The building still stands and is located just south of the C&EI underpass on the east side of the highway as one enters Crete. (Courtesy Carol Triebold.)

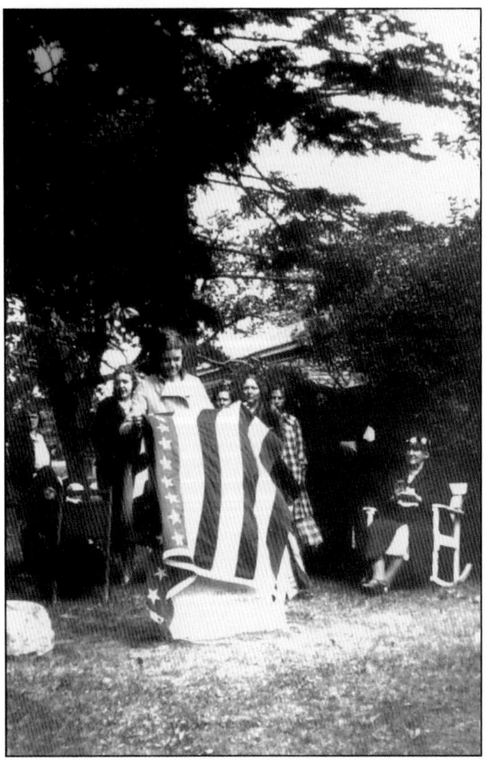

Dedication ceremonies took place in 1936 for a marker to commemorate Hubbard's Trail, much of which Illinois Route 1 and the Dixie Highway follow closely north of Danville. The marker was placed in front of the George Kester home in Crete by the Will County Centennial Committee. (Courtesy Carol Triebold.)

George Sharp's Texaco service station was located in Crete on the northwest corner of Fifth and Main Streets, the Dixie Highway. Besides selling gas and oil, the station had an exterior lift to facilitate oil changes and repair work. The station was enlarged over the years, and the building now houses the Cone Cottage ice-cream shop. (Courtesy Carol Triebold.)

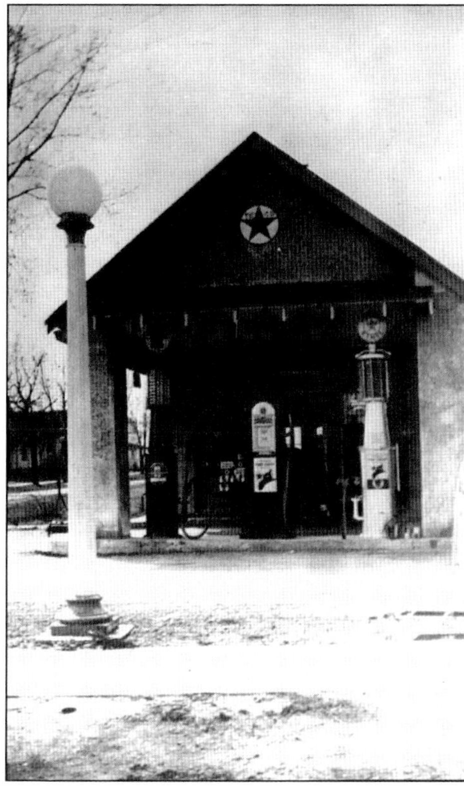

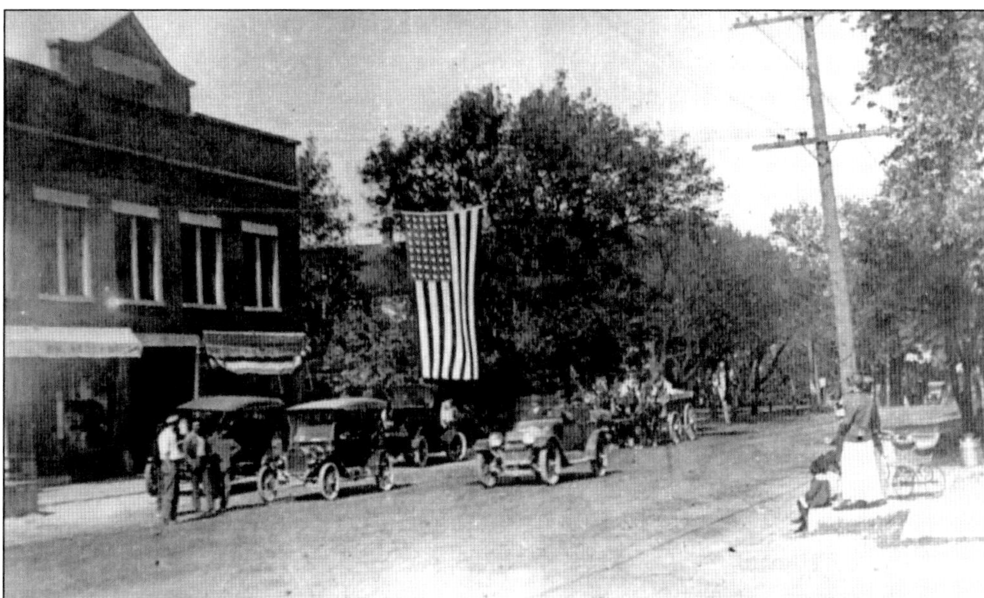

In October 1915, organizers and boosters of the Dixie Highway planned a trip to inspect the conditions of the newly routed highway in the states it passed through. The trip started in Chicago on October 9 and passed through Crete that same morning. Part of the motoring contingent is shown traveling south on the yet unpaved stretch of Main Street, just north of Exchange Street. (Courtesy Carol Triebold.)

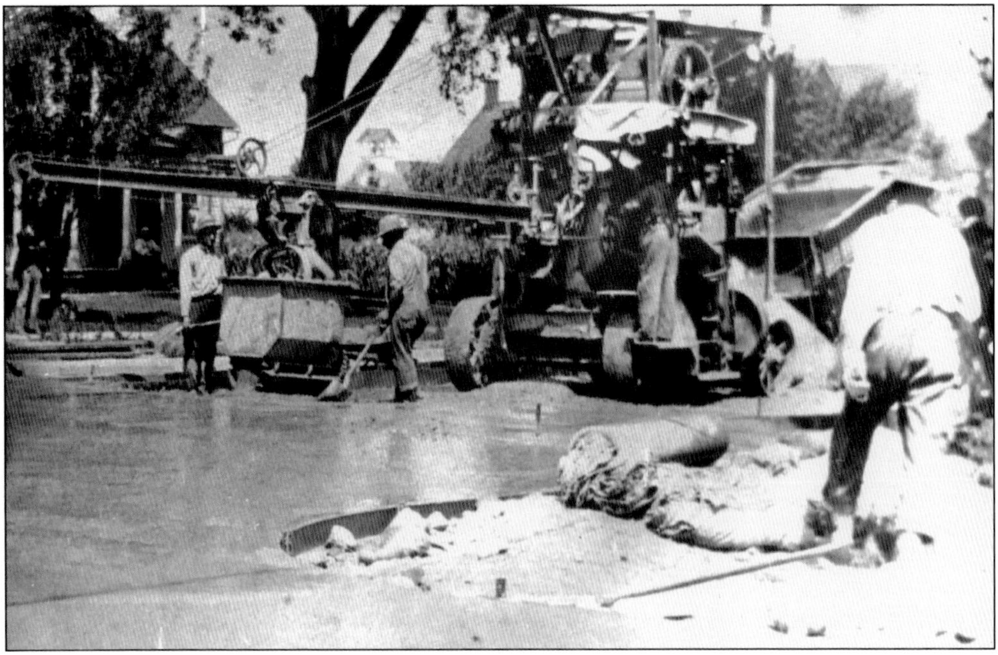

Paving work on Main Street and the Dixie Highway was completed in Crete by June 1920. Work was completed by Joe Orr's Chicago Heights Coal Company and included a 30-foot-wide concrete pavement with curbs and gutters in town. Here workers are shown ready to unload another fresh batch of cement on the highway. (Courtesy Carol Triebold.)

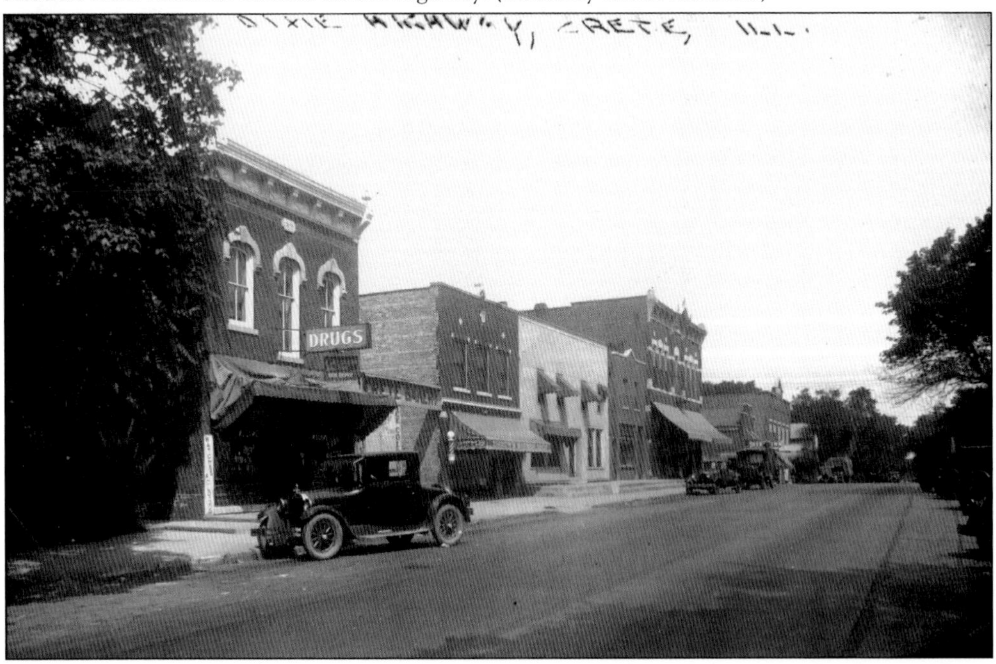

Traffic was light on the new pavement when this view was taken of the west side of Main Street, the Dixie Highway, looking north toward Exchange Street in Crete. Rohe Brothers operated its drugstore in the building on the left while to the right are the Crete Bakery and Zirzow's Barber Shop. (Courtesy Carol Triebold.)

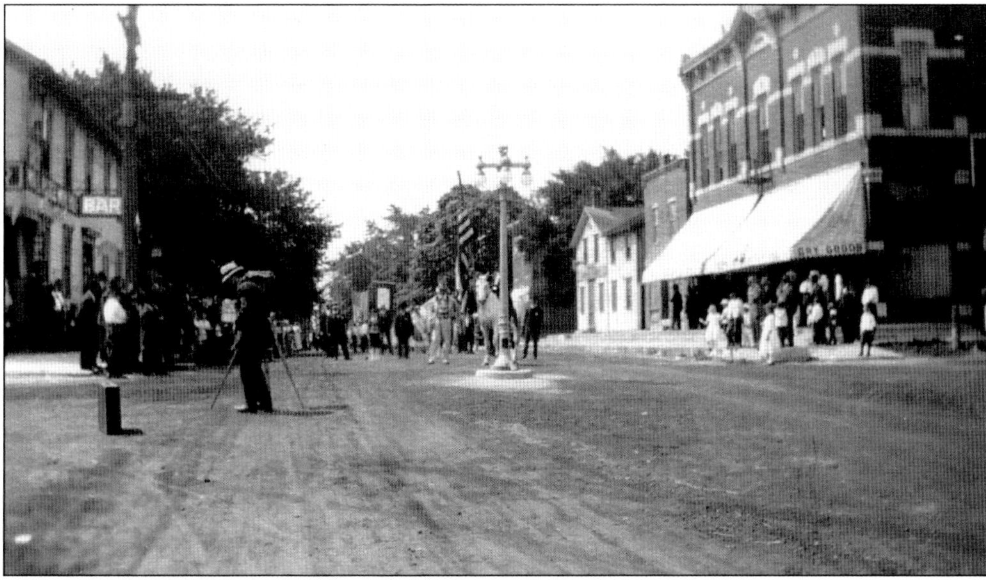

Uncle Sam is set to lead a parade north on Main Street, the Dixie Highway, at Exchange Street in Crete. The decorative light post was installed in the center of the intersection at the end of 1916 and was a donation of merchant Gus Wilkening. The fixture was an attractive addition to the intersection but was soon damaged by passing automobiles and was later relocated. (Courtesy Carol Triebold.)

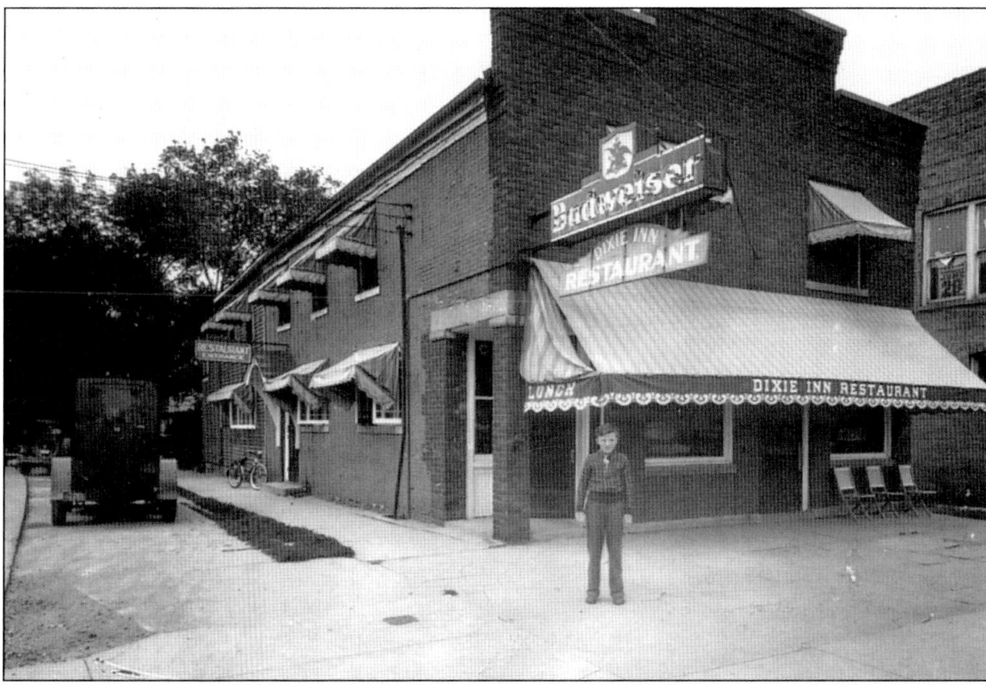

By the 1930s, the building on the northwest corner of Main and Exchange Streets in Crete was known as the Dixie Inn, a tavern and restaurant, owned by Gus Krebs. The building formerly housed the William Seggebruch Tavern and has since been torn town. A newer building housing Matt's Printing is now located on the site. (Courtesy Carol Triebold.)

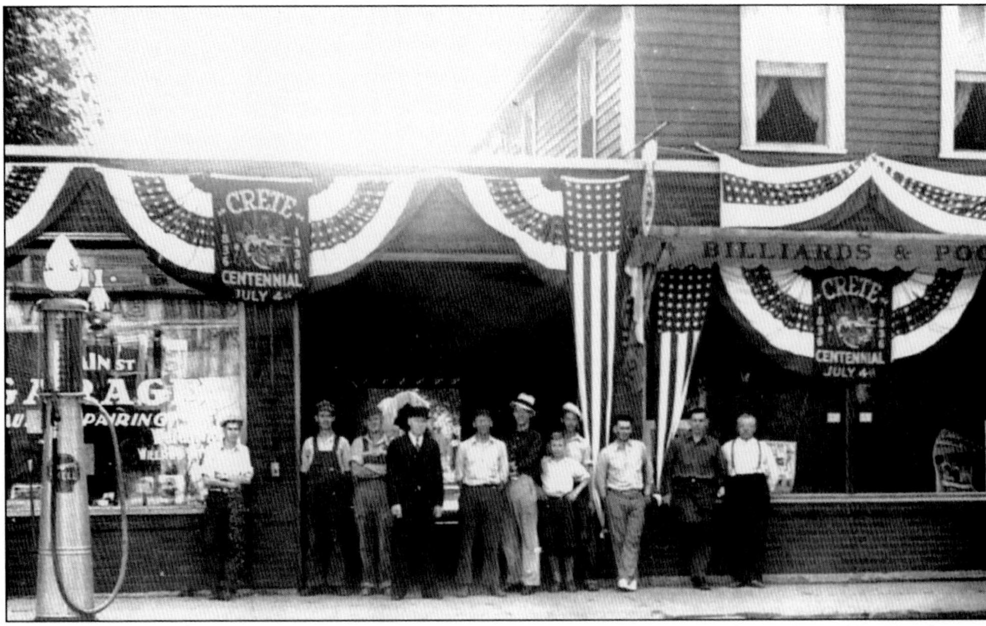

Adolph Rinne owned the Main Street Garage (1373 Main Street), and Hulda Buchanan owned the adjacent billiards and pool hall when this photograph was taken during Crete's 1936 centennial celebration, which marked the 100th anniversary of Willard Wood's arrival in the area. Pictured are, from left to right, Kenneth Muehring, F. R. Devert, Al Grimm, ? Farar, Adolph Rinne, C. Vernon Boehner, H. Otto Luehrs, Ervin Paul, ? Kehl, Roy Devert, and ? Stier. (Courtesy Carol Triebold.)

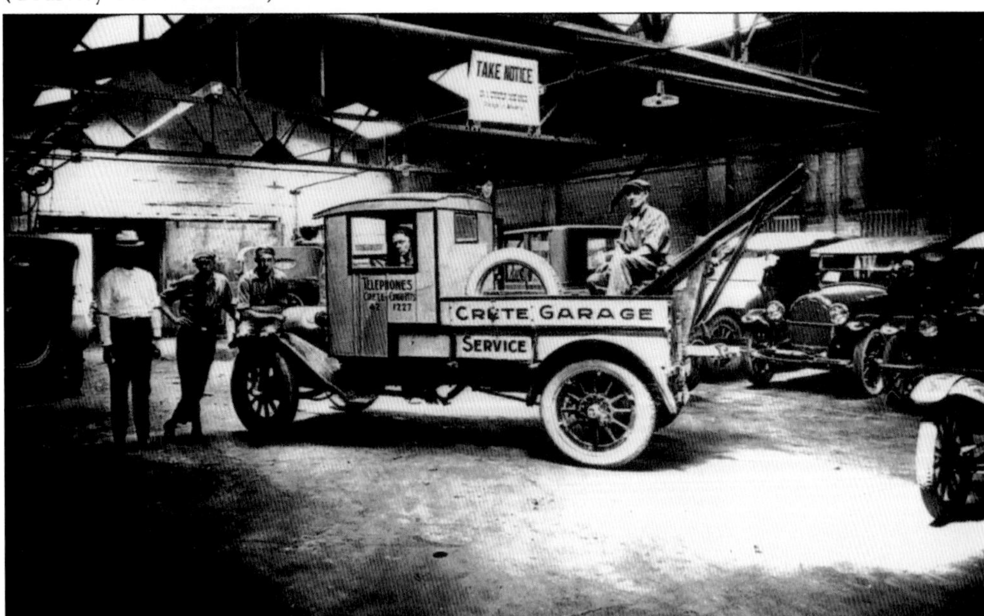

William Seggebruch operated a garage and automobile dealership at 1378 Main Street in Crete. Seggebruch's Crete Garage provided full service, which included gas, oil, repairs, towing, and indoor storage. This view of the interior of the building shows employees proudly posing with the garage's wrecker, or tow truck. (Courtesy Carol Triebold.)

WLS Radio, established in 1924 by Sears, Roebuck and Company, located its broadcast transmission towers on the Dixie Highway just south of Crete. Known as the Prairie Farmer station, it was popular in the rural Midwest and was home to the WLS *Barn Dance* radio program. The station's call letters reflected Sears's position as the "World's Largest Store." The Crete broadcast facilities closed in 1938. (Courtesy Carol Triebold.)

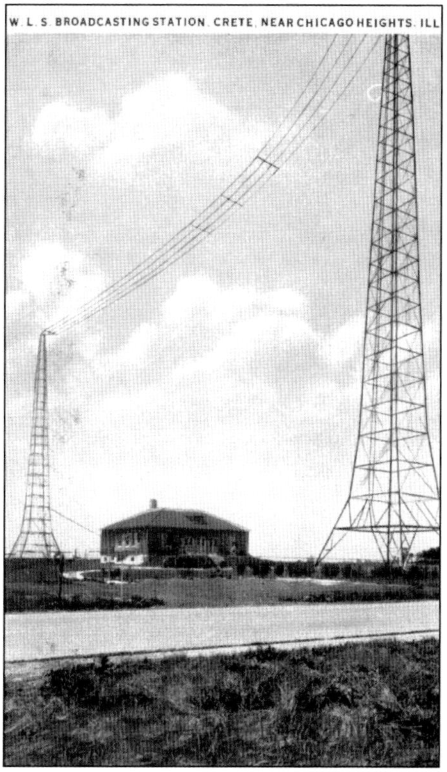

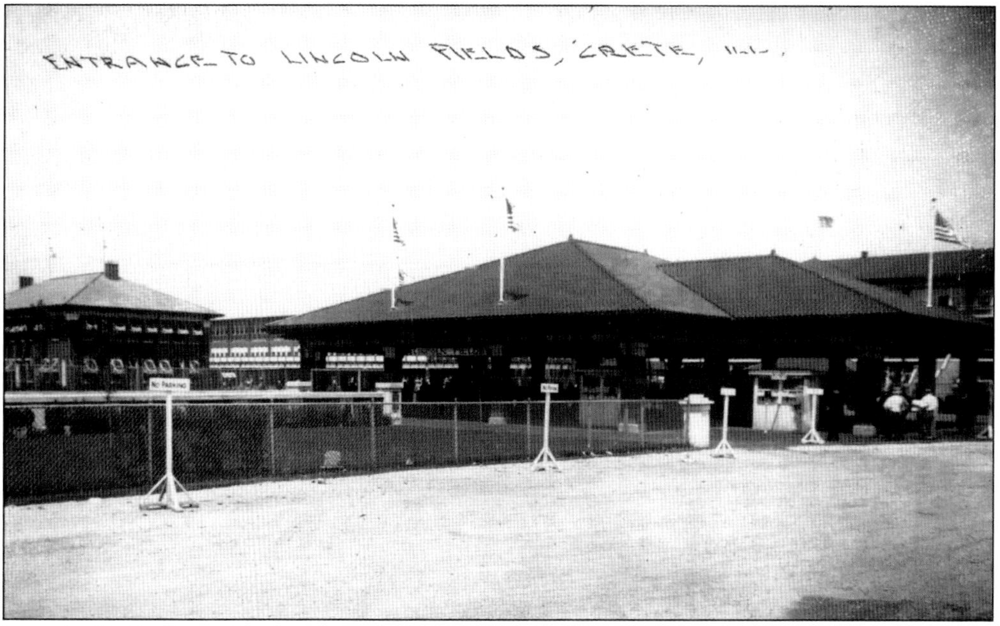

Lincoln Fields Race Track was a popular destination on the Dixie Highway south of Crete. The track opened on August 9, 1926, for its inaugural meet of races. In addition to the Dixie Highway, race fans also traveled to the racecourse along a spur line of the C&EI. Passengers were discharged on the west side of the highway and walked through an underpass, which led to this entry to track grounds. (Courtesy Carol Triebold.)

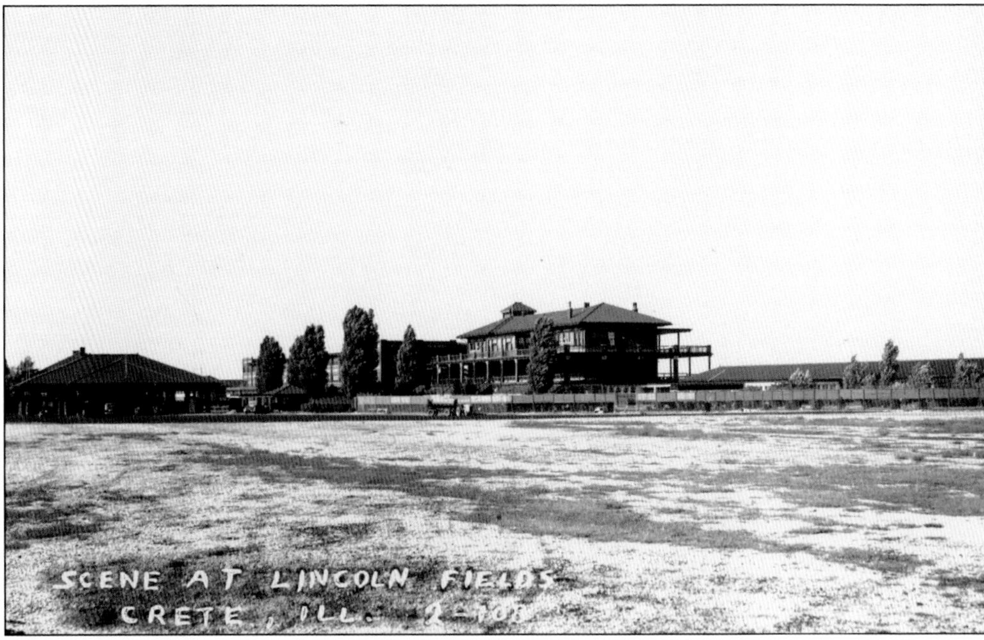

The original clubhouse building at Lincoln Fields in Crete (center) was the scene of a disastrous fire on January 30, 1952. The clubhouse was declared a total loss, and a new building was completed in time for the start of the racing season in May 1954. The Balmoral Jockey Club purchased Lincoln Fields in 1955, and the track has been known as Balmoral ever since. (Courtesy Carol Triebold.)

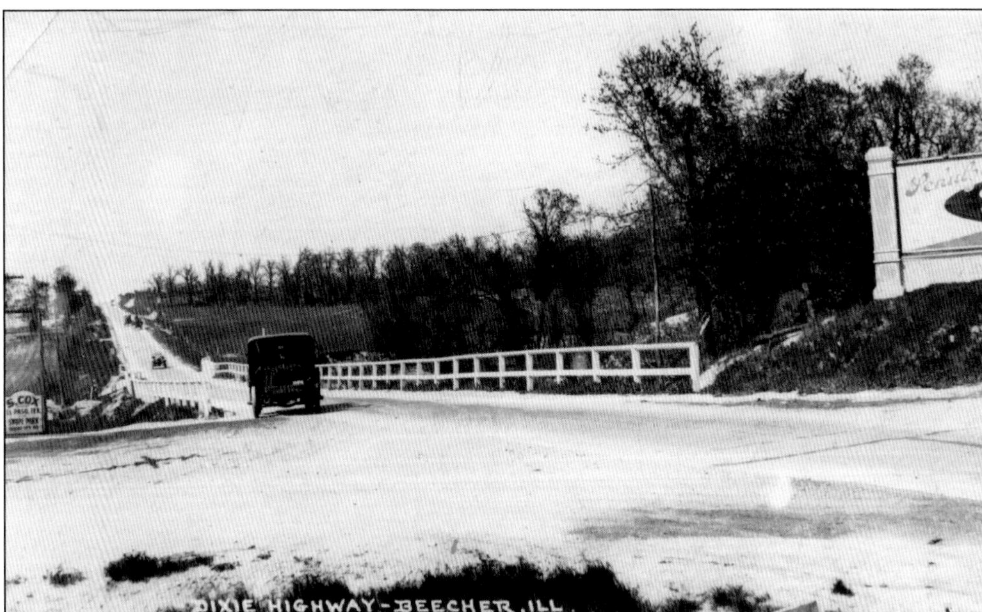

This view looks north on the Dixie Highway from Goodenow Road near Goodenow. With the exception of the vintage vehicles and signage, this roadway scene looks quite similar today. The State of Illinois has future plans to extensively reconfigure the intersection where the Dixie Highway, Goodenow Road, and Illinois Route 394 meet. (Courtesy Carol Triebold.)

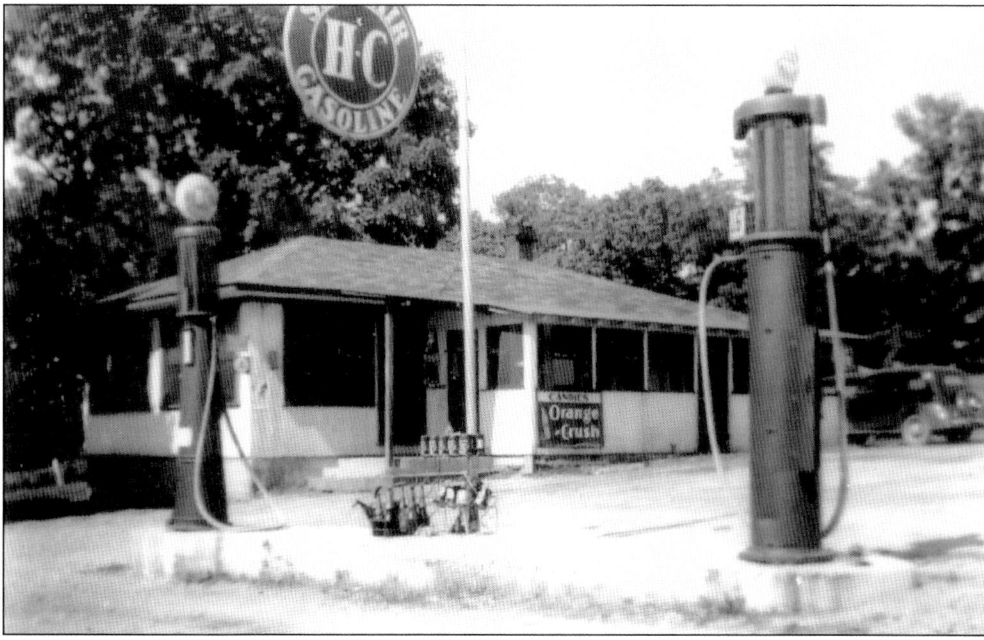

The Collins Sinclair gas station was located on the northwest corner of the Dixie Highway and Goodenow Road. Typical of gas stations at the time, oil and watering cans were kept at the ready near the gas pumps to allow for quick service of customers' vehicles. Today this site near Goodenow is vacant, but a hot dog stand had been located here for years. (Courtesy Carol Triebold.)

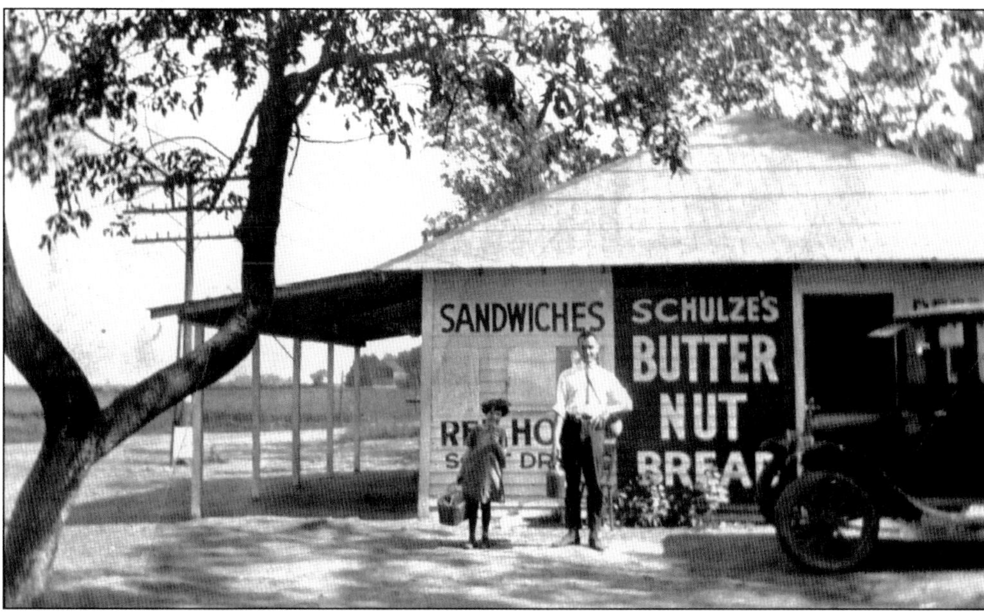

Henry Kregel owned this small eatery on the east side of the Dixie Highway south of Church Road in Beecher. Fare consisted of plate lunches, barbecue sandwiches, and chicken dinners, which were advertised as "something to crow about." Kregel also sold Shell gasoline from two pumps next to the roadway. Erich Ohlendorf and Helen Hall are pictured. (Courtesy Marilyn Ohlendorf.)

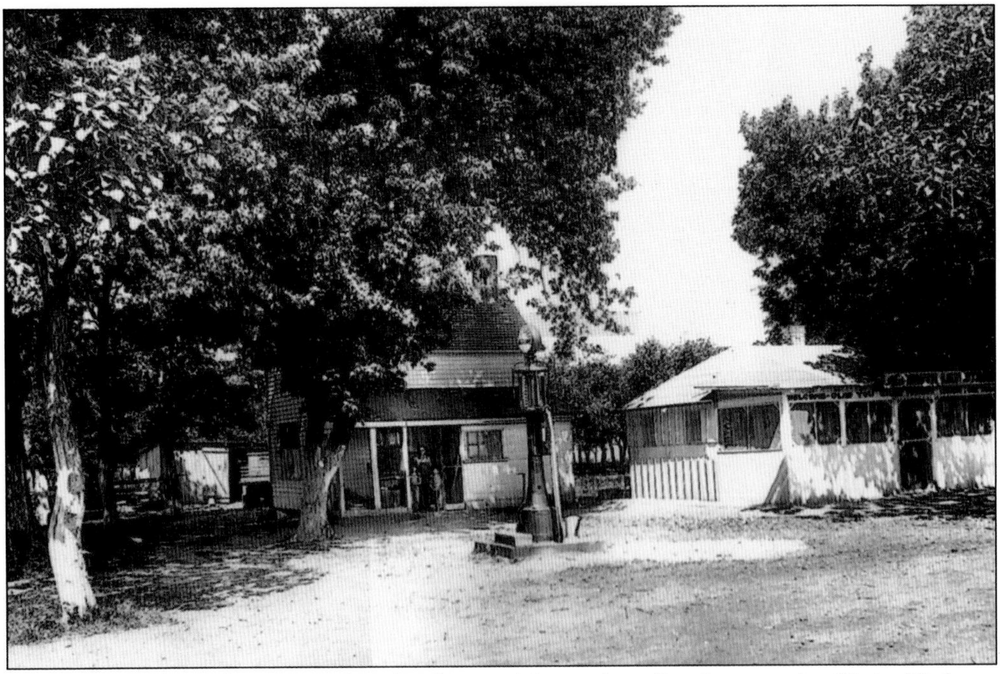

The White Lily Café was another lunch stand located in Beecher on the Dixie Highway south of the Kregel place. Owner Charles Whitehead named the place for his wife, Lillian. As diners entered the café they were greeted with friendly signage above the doorway that read, "Welcome—Glad You Came—Hurry Back—We Want Your Business." (Courtesy Beecher Historical Society.)

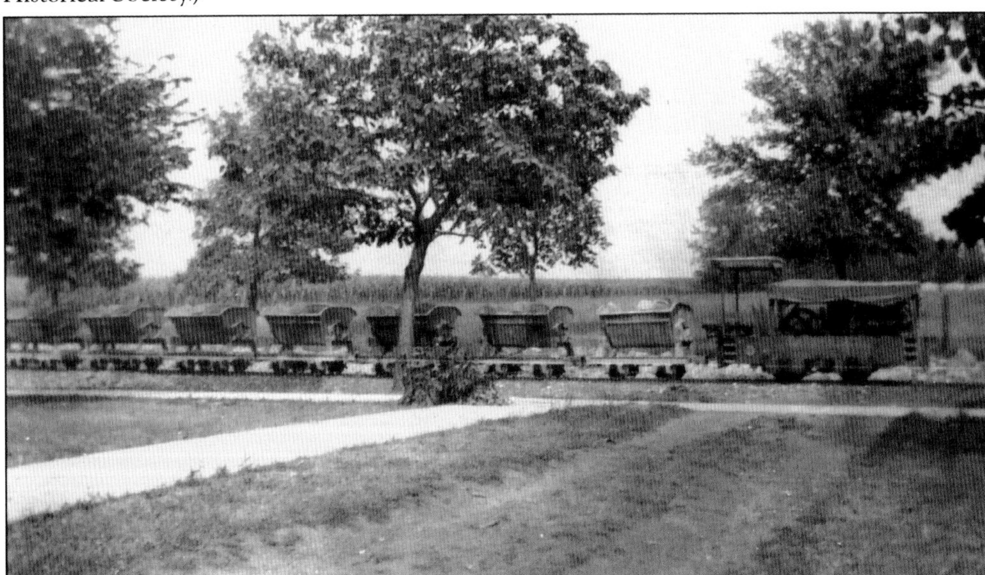

Paving on the Dixie Highway was completed through Beecher in May 1921. A stretch north of town was some of the last of the highway to be paved in Illinois and was officially opened to traffic on June 2. Paving contractors installed light-rail tracks along the road's right-of-way and used this small engine and hopper cars to transport raw materials to the job site. (Courtesy Beecher Historical Society.)

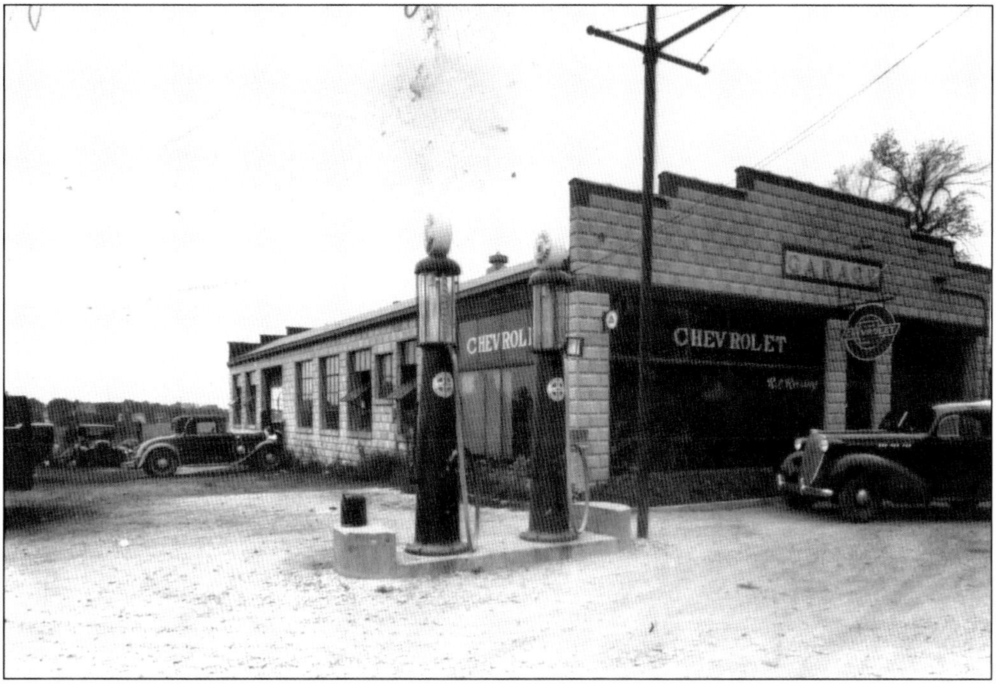

Roy Reising opened his Chevrolet dealership in Beecher in 1929. He moved operations into this building at 700 Dixie Highway in 1936. The structure and 20 automobiles in it were destroyed by fire in March 1938. Reising rebuilt, and the dealership is still in business today on the Dixie Highway. (Courtesy Reising Chevrolet.)

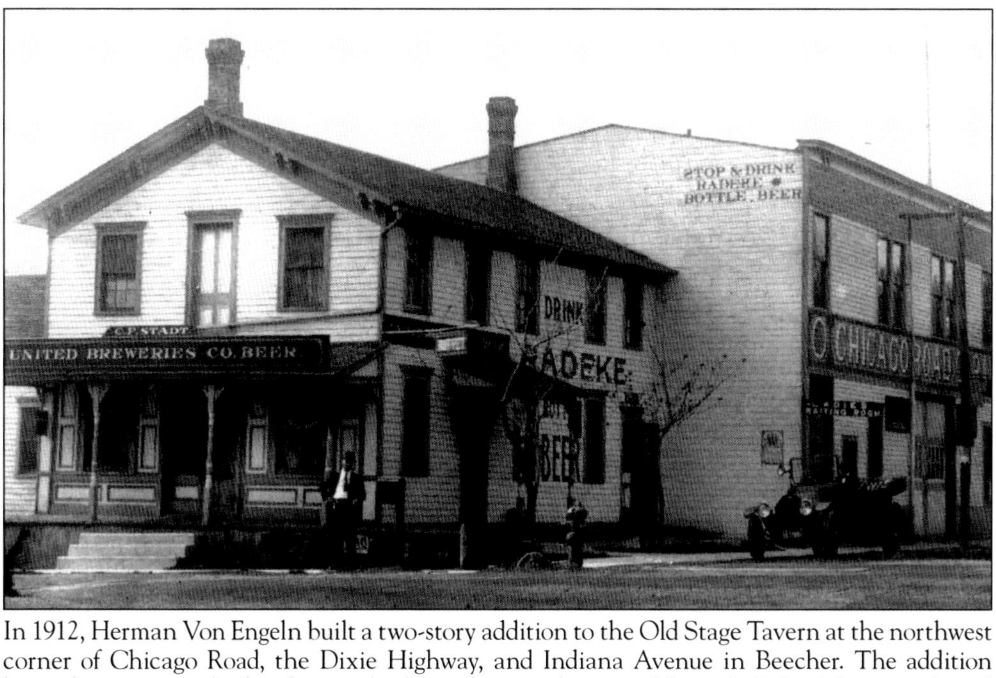

In 1912, Herman Von Engeln built a two-story addition to the Old Stage Tavern at the northwest corner of Chicago Road, the Dixie Highway, and Indiana Avenue in Beecher. The addition housed a garage on the first floor and a dance floor on the second floor. C. F. Stadt later purchased the property. The original structure dates from 1870. (Courtesy Beecher Historical Society.)

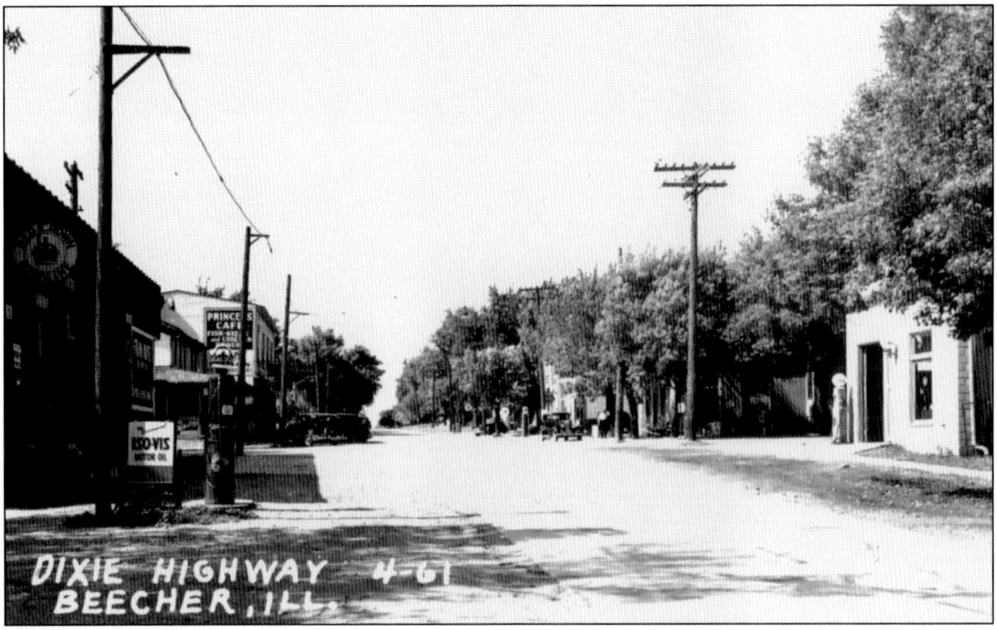

Traffic was quite light in Beecher when this photograph from the late 1930s was taken of the Dixie Highway looking north at the intersection with Indiana Avenue. At left are Frank Hunte's general store and the Old Stage Tavern, which was now known as the Princess Café. At right is Langreder's Service Garage, which was built in 1928. (Courtesy Beecher Historical Society.)

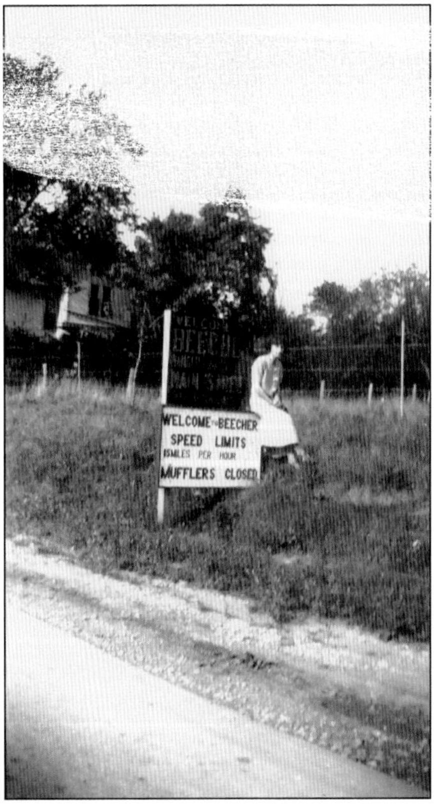

One of Beecher's fair maidens poses at the village's welcome sign at the south end of town. Motorists entering the community are advised the speed limit is 15 miles per hour and mufflers are to be "closed." The upper sign warns of a crossroad ahead, and an arrow points visitors to the "Main Street 4 Blocks West." (Courtesy Beecher Historical Society.)

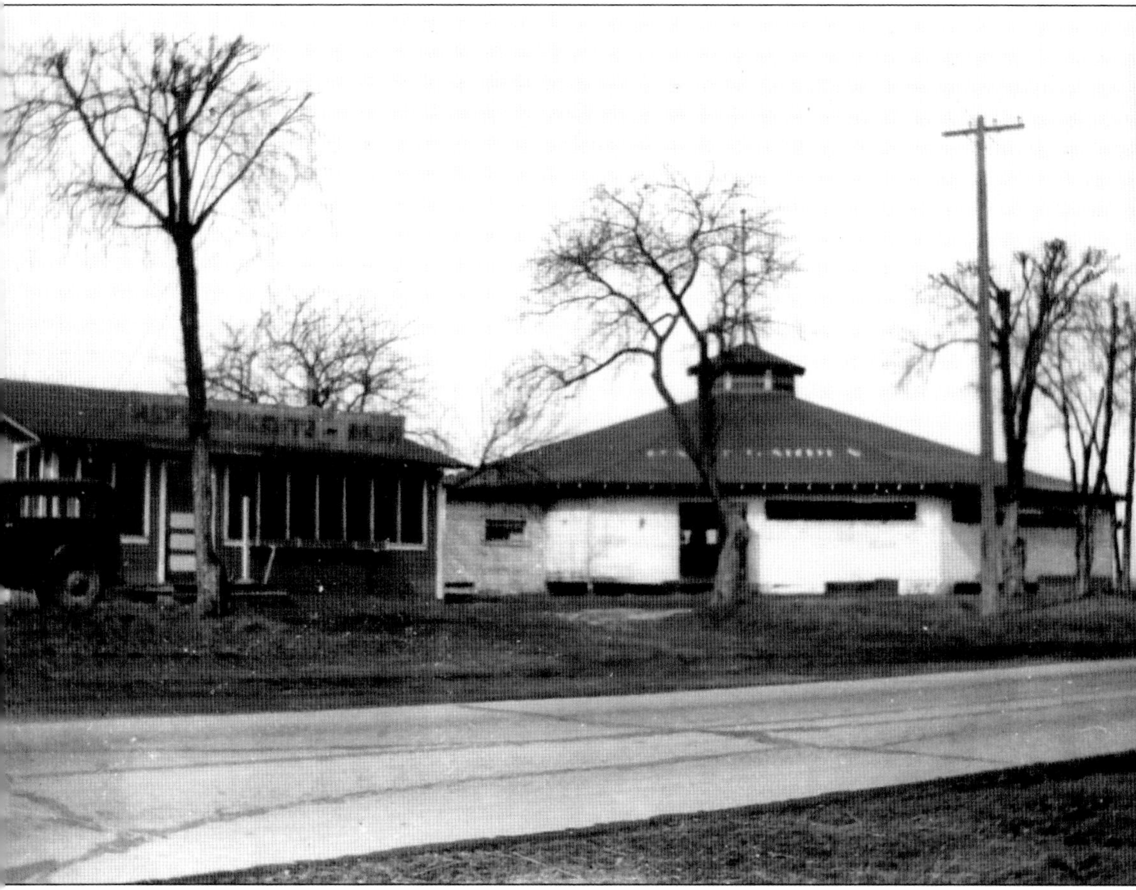

The Palm Gardens dance pavilion was located outside Beecher on the west side of the Dixie Highway just south of Kentucky Road on part of the Fiene farm. The dance hall was initially run by a man named Davis from Chicago Heights and was then operated by George Bydalek and his wife, Mary. The dance hall was in business from 1925 to 1936 or 1937, when it was torn down. (Courtesy Sylvia Unruh.)

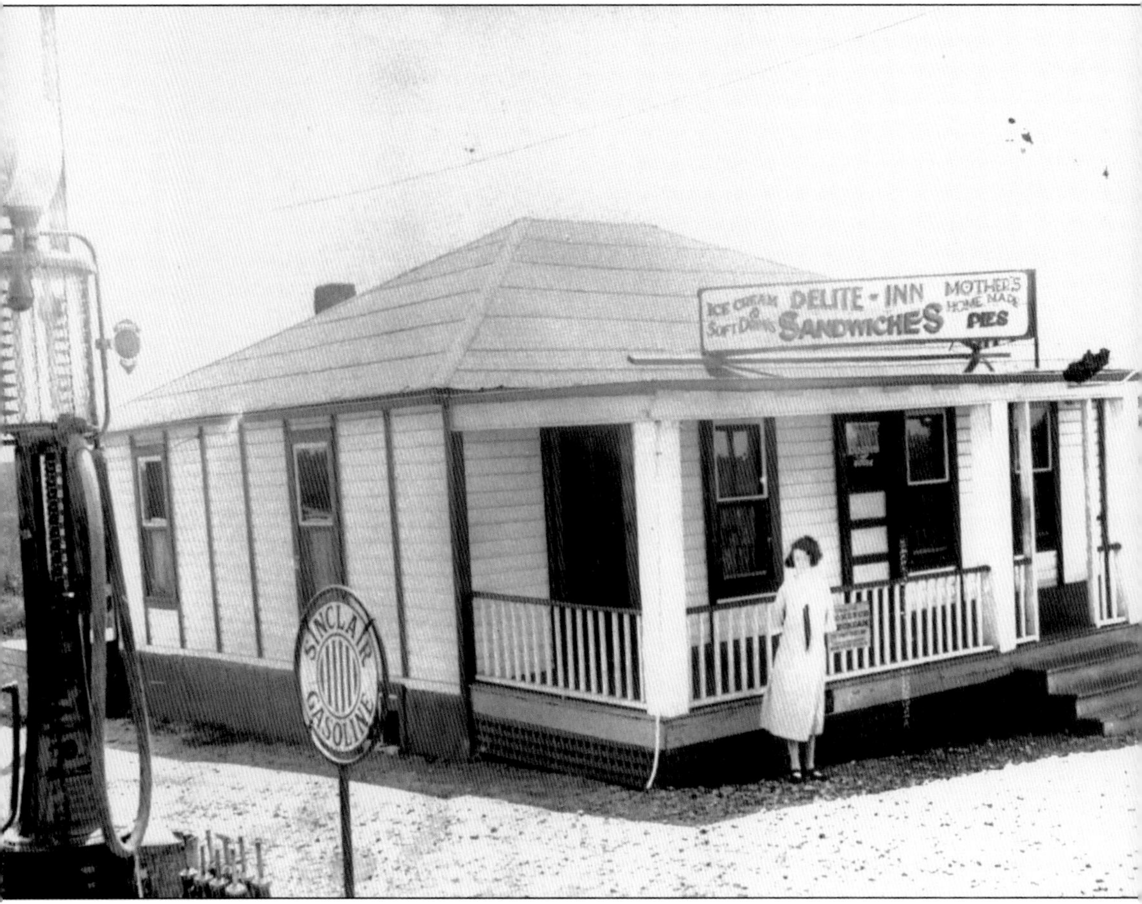

Alice Leder stands in front of the Delite Inn that she operated with her husband Frank. Ice cream, soft drinks, sandwiches, and "mother's home made pies" were served inside the restaurant, and Sinclair brand gas and oils were sold outside. The building still stands on the east side of the highway at the Delite Inn Road south of Beecher. (Courtesy Janet Nelson.)

Three

Kankakee County

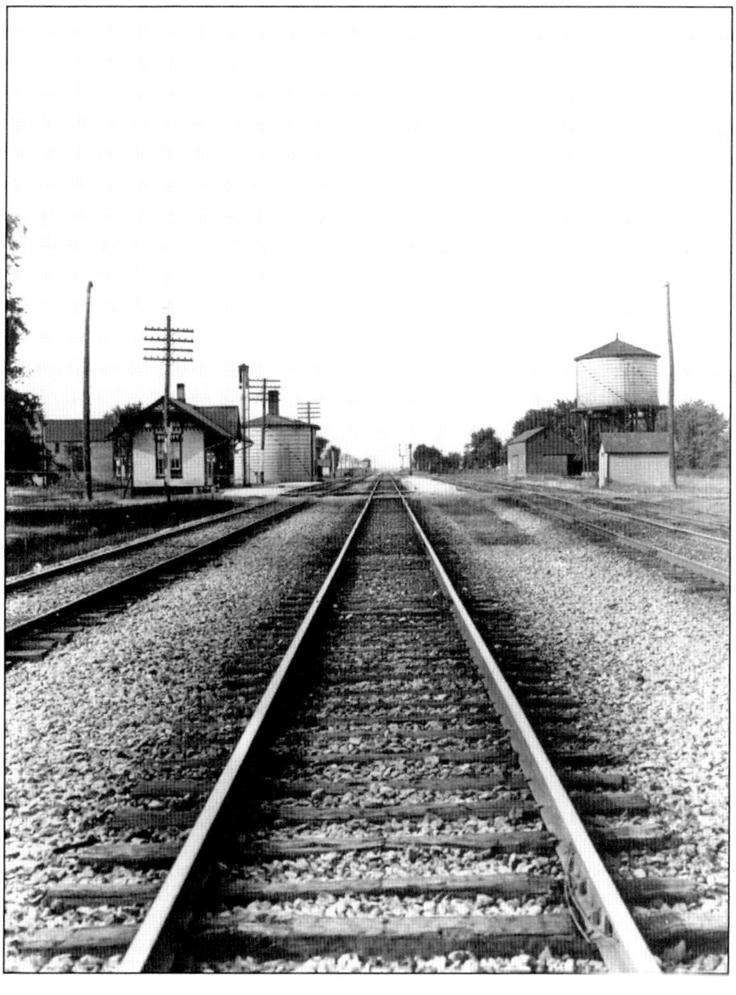

A small hamlet, Sollitt is located west of Illinois Route 1, the Dixie Highway, at the border of Will and Kankakee Counties and was once a stop on the C&EI. The depot is long gone, and Sollitt Tap, a popular stop for motorcycle enthusiasts, is the sole business operating in the community. (Courtesy Esther Kimberly.)

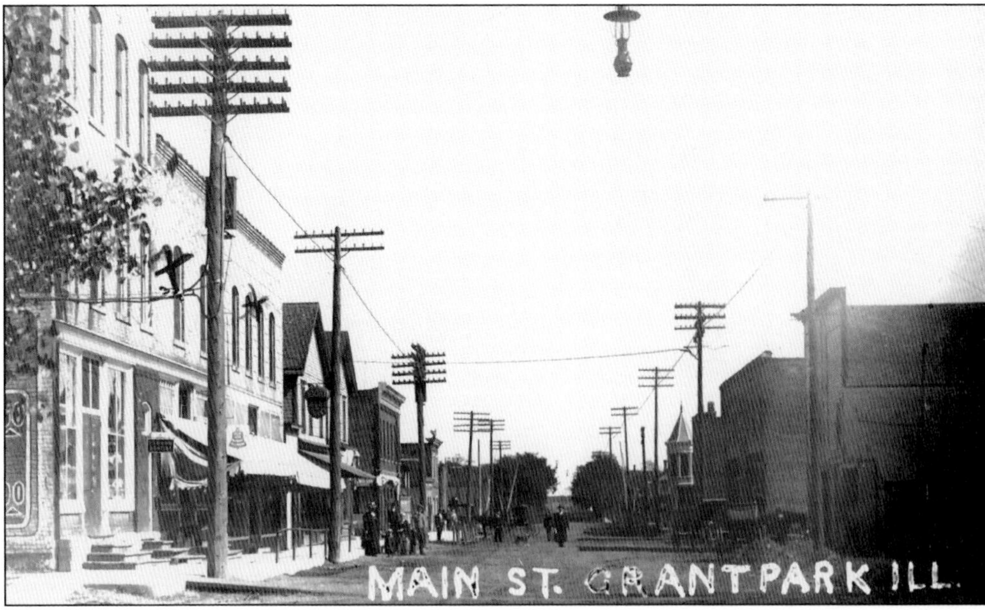

The title identifies this postcard as the "main" street in Grant Park. The view looks east on Taylor Street from Maple Street toward the intersection of Main Street and the railroad crossing just beyond. The original route of the Dixie Highway traveled along this stretch of roadway until 1920 when paving on the current alignment of the highway was completed. (Courtesy Nathan Huegel.)

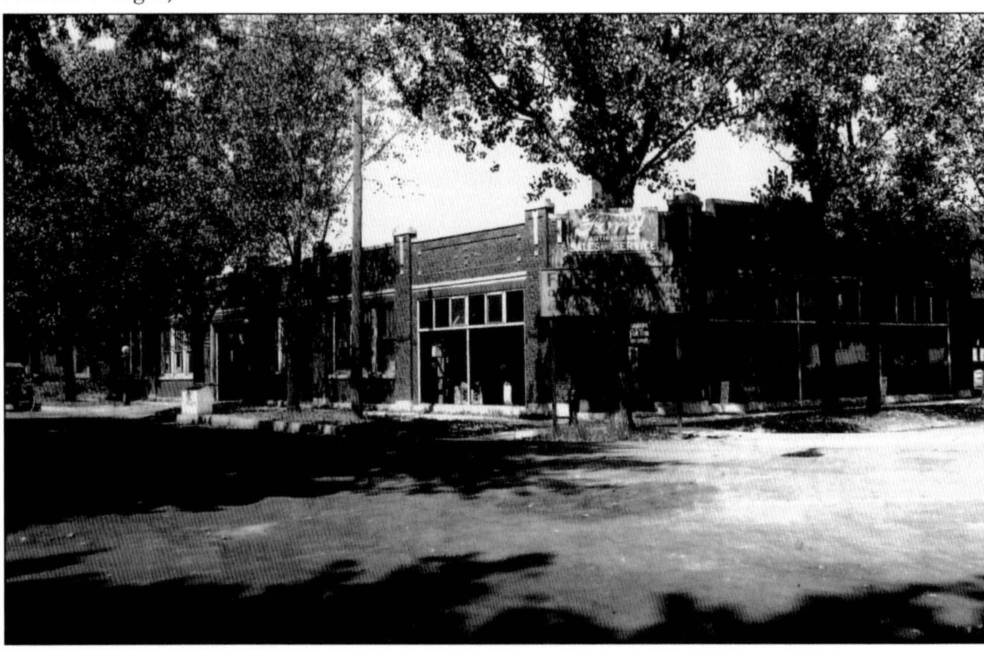

Located on the corner of Taylor and Maple Streets, the Grant Park Motor Company boasted being "the Finest Garage on the Dixie Highway." The early Ford dealership was sold to Henry C. Wehrmann in October 1921 and was one of three automobile dealerships he had in towns along the Dixie Highway. The others were located in Goodenow and Chicago Heights. (Courtesy Grant Park Historical Society.)

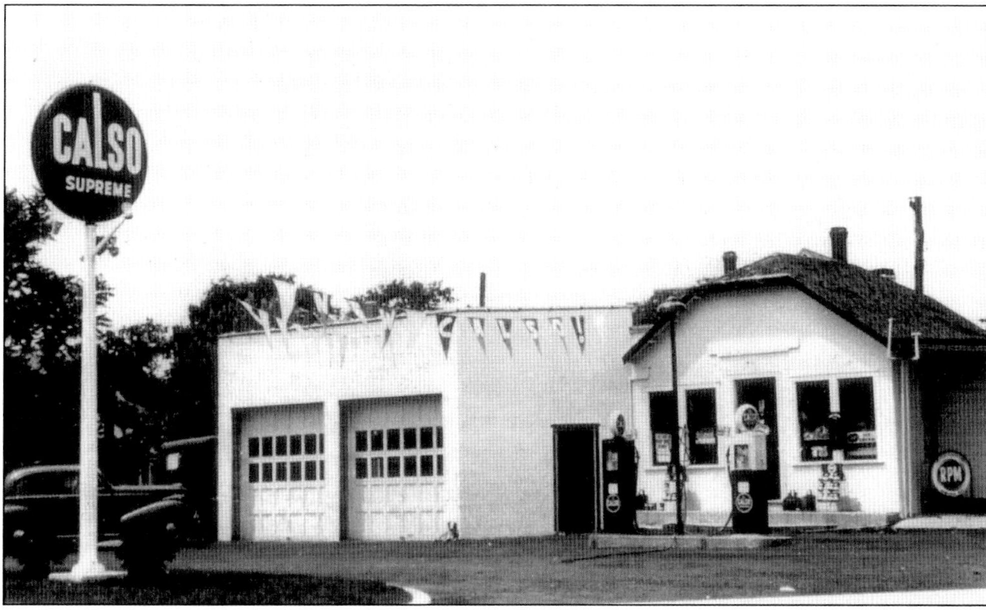

The Calso station was located in Grant Park at the corner of Main Street, the Dixie Highway, and Alma Street. The station sold the products of Standard Oil of California and all company locations in the eastern United States were branded Calso, while company stations in the West were known under the Chevron name. In 1958, all stations were converted to the Chevron brand. The company no longer markets in Illinois. (Courtesy Landis Wehrmann.)

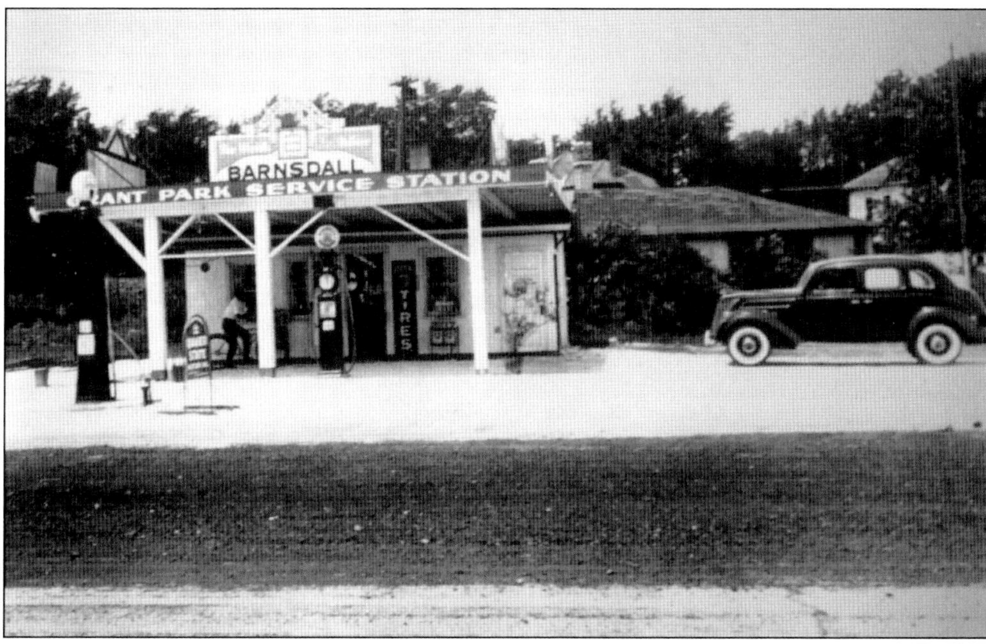

The Grant Park Service Station was located on the curve of Main Street, the Dixie Highway, on the south end of town. The station sold Barnsdall Oil products. The Oklahoma-based company proclaimed itself "the World's Finest Refiner" and also touted the slogan "Be Square to Your Motor" on station signage. Today a Shell gas station and convenience store occupy the site. (Courtesy Vern Koelling.)

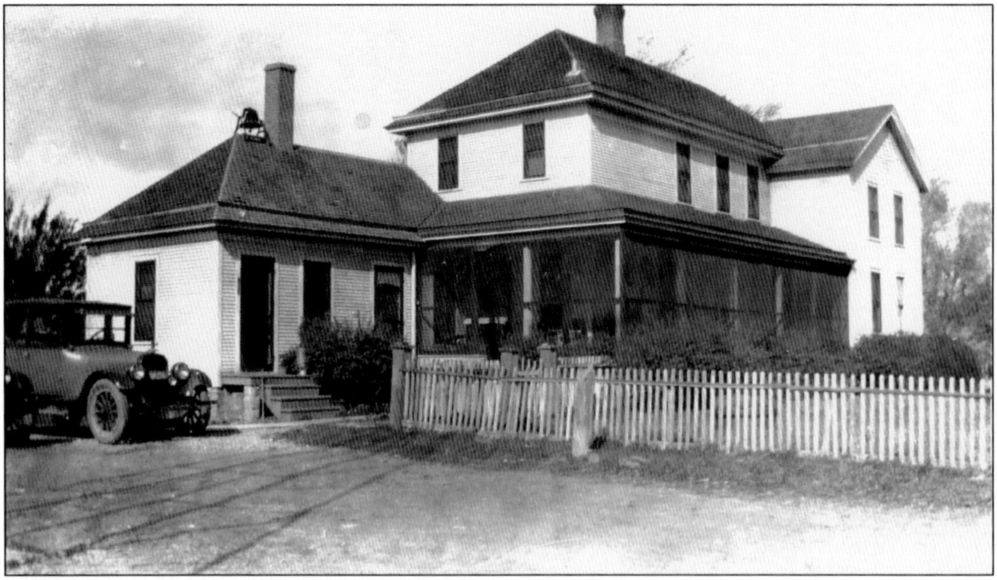

Farmhouses such as this one dot the whole route of the Dixie Highway in Illinois. Area pioneer Sam Beebe built the home, located on the west side of the highway between Grant Park and Momence, in about 1850. Members of the Bradley family lived in the home for almost seven decades. The home has been greatly enlarged over the years by the Schriefer family and now has a distinctly Victorian look. (Courtesy Harold and Amelia Schriefer.)

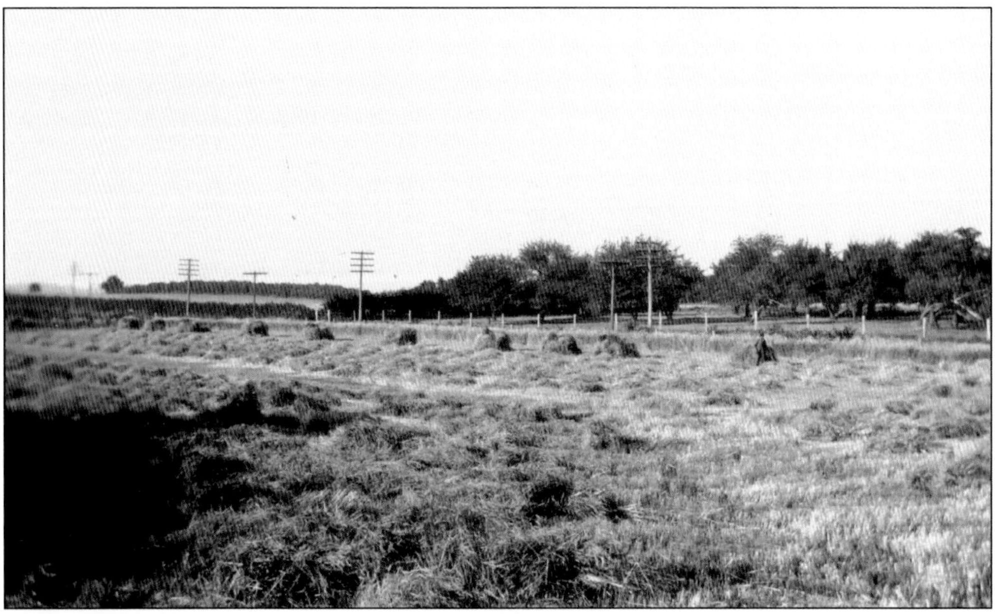

Agriculture has and still does dominate the landscape while "drivin' the Dixie." In this pastoral view, shocks of wheat await threshing in the fields of the Bradley farm located between Grant Park and Momence. The Dixie Highway parallels the telephone poles in the background of this photograph from 1918. (Courtesy Harold and Amelia Schriefer.)

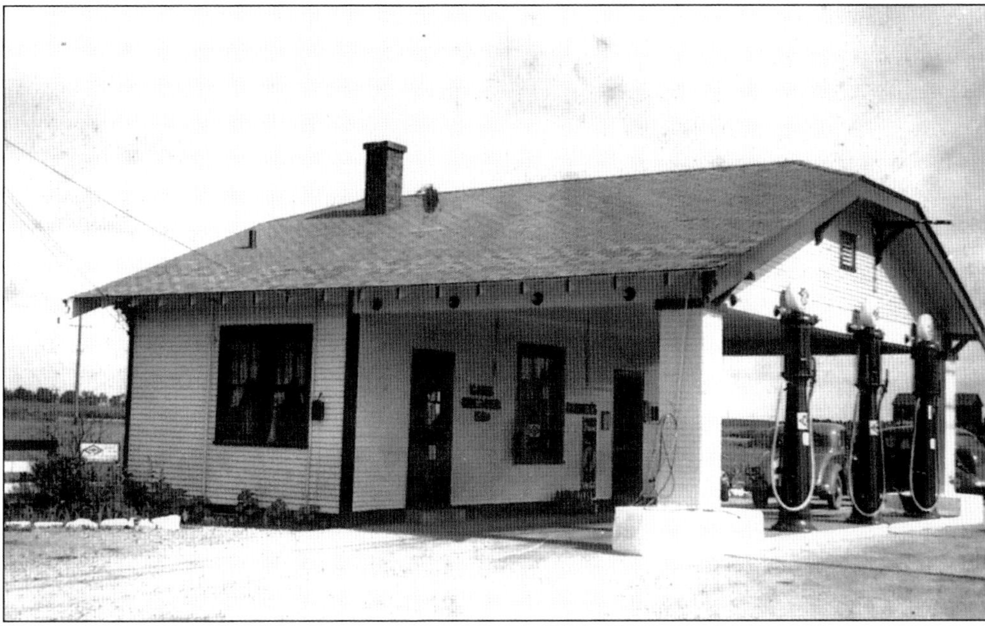

Barney's Service Station was located on the north end of the Dixie Highway in Momence. The station was located on the edge of town for years and was owned by Barney Broulette. He sold DX brand gasoline, and when this photograph was taken, a grease job cost 50¢. Today a modern gas station and convenience store occupies the site. (Courtesy Edward Chipman Public Library/Kevin McNulty.)

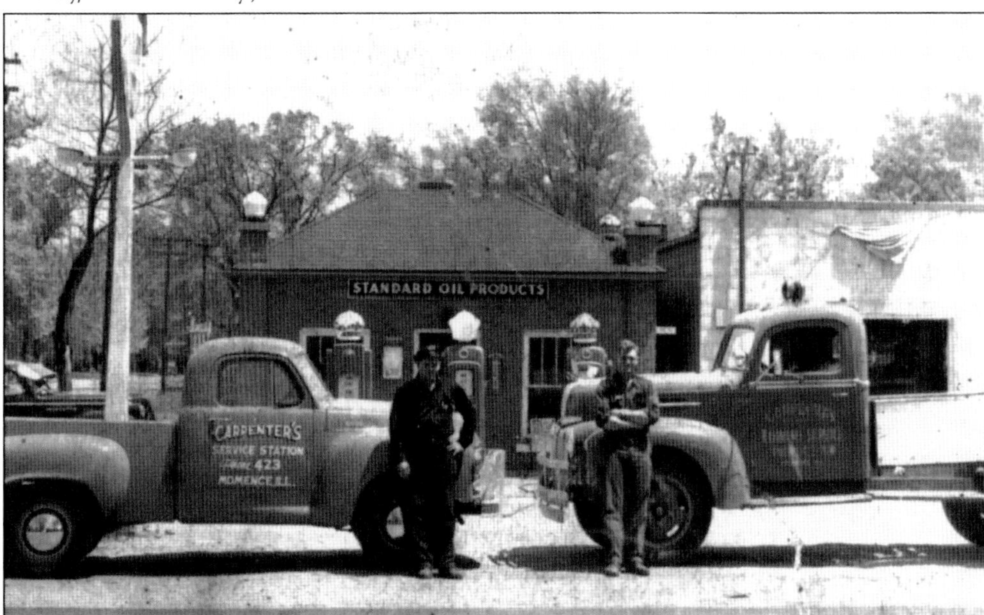

Robert Carpenter, left, and DeWitt Carpenter stand in front of the Standard Oil service station they operated in Momence on the Dixie Highway at the southeast corner of Sixth Street. Although the photograph was taken in the 1950s, the station dates to the 1920s and is a nice example of early gas station architecture. The building still stands and is a used car dealership. (Courtesy Vicki Carpenter Sommer/Kevin McNulty.)

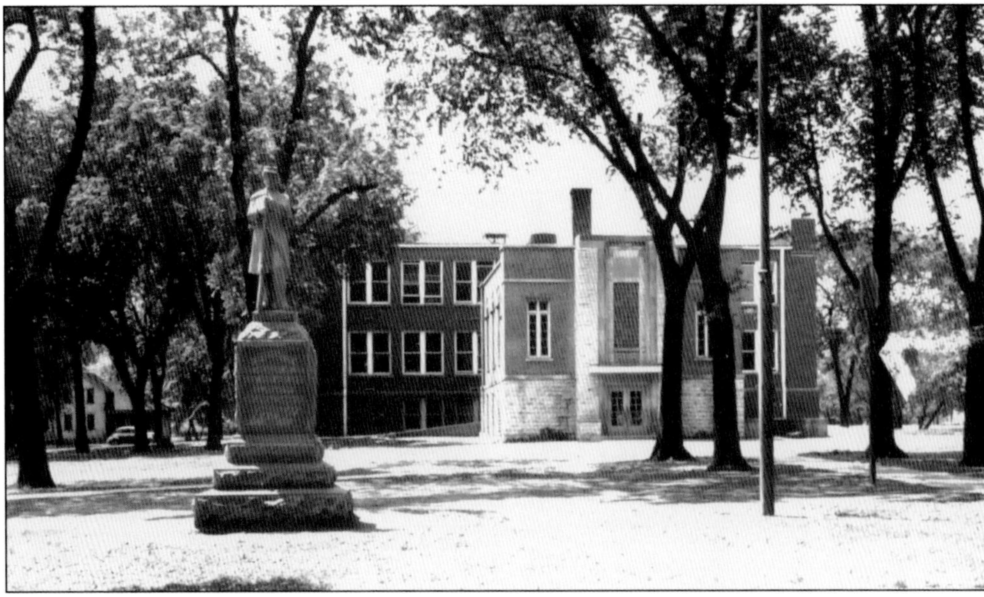

The statue of a Civil War soldier has stood in Momence as a sentinel overlooking the Dixie Highway for over 100 years. The statue was dedicated on July 5, 1909, and was erected by the Women's Relief Corps, an organization formed during the war to furnish nursing supplies, clothing, and other necessities to ill and wounded northern soldiers. (Courtesy Edward Chipman Public Library/Kevin McNulty.)

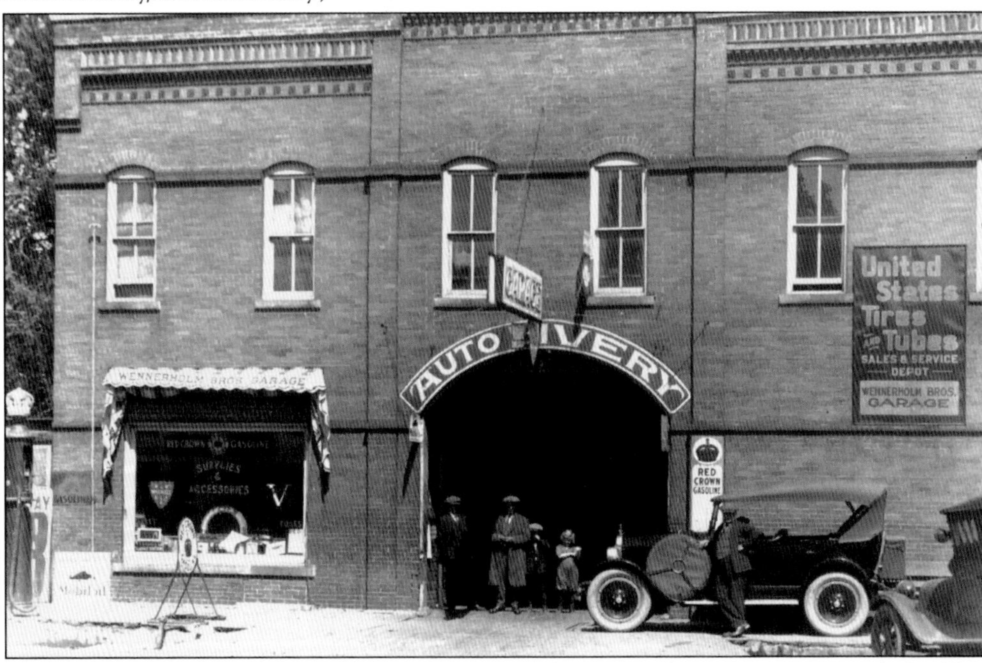

Wennerholm Brothers operated this garage at 100 North Dixie Highway in Momence. The building was constructed in 1885 with corbeled brickwork. Its facade touts a variety of products including Standard Red Crown gasoline, Mobil Oil products, Willard batteries, and United States tires and tubes. Pictured are, from left to right, August Wennerholm, Nelson Osby, Fred Madsen, Ellen Madsen, and Gus Wennerholm. (Courtesy Edward Chipman Public Library/Kevin McNulty.)

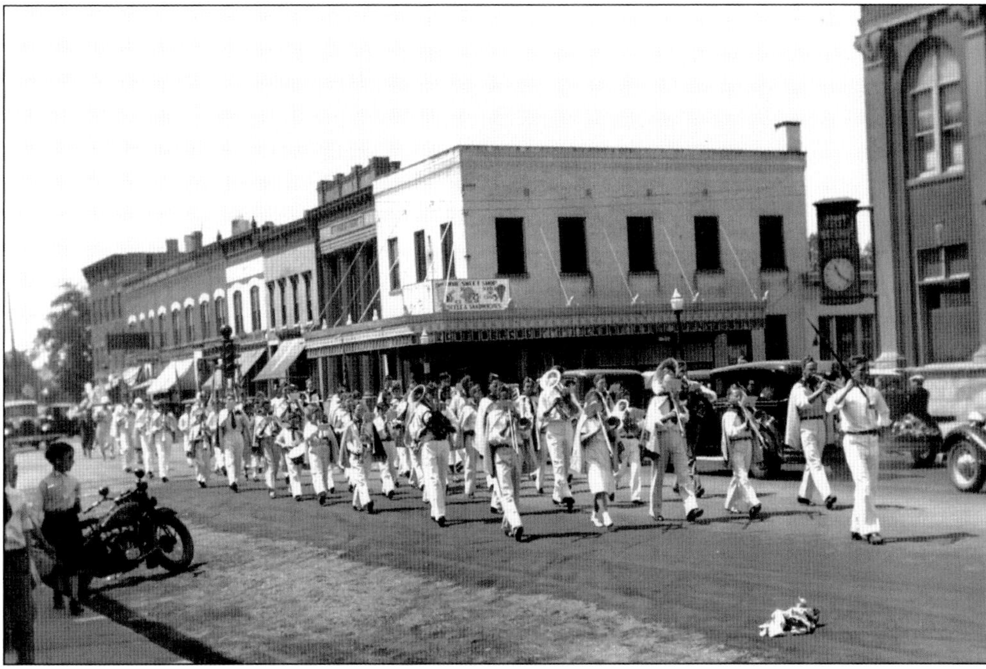

A school band marches north on the Dixie Highway at Washington Street in this interesting view of Momence from the 1930s. The 1914 classic revival–style Parish Bank building and the Dixie Sweet Shop, located at the southwest corner of the intersection, can be seen in the background. The sweet shop structure, built in 1921, was constructed of enameled brick produced locally by the Tiffany Brick Company and has since been demolished. (Courtesy Edward Chipman Public Library/Kevin McNulty.)

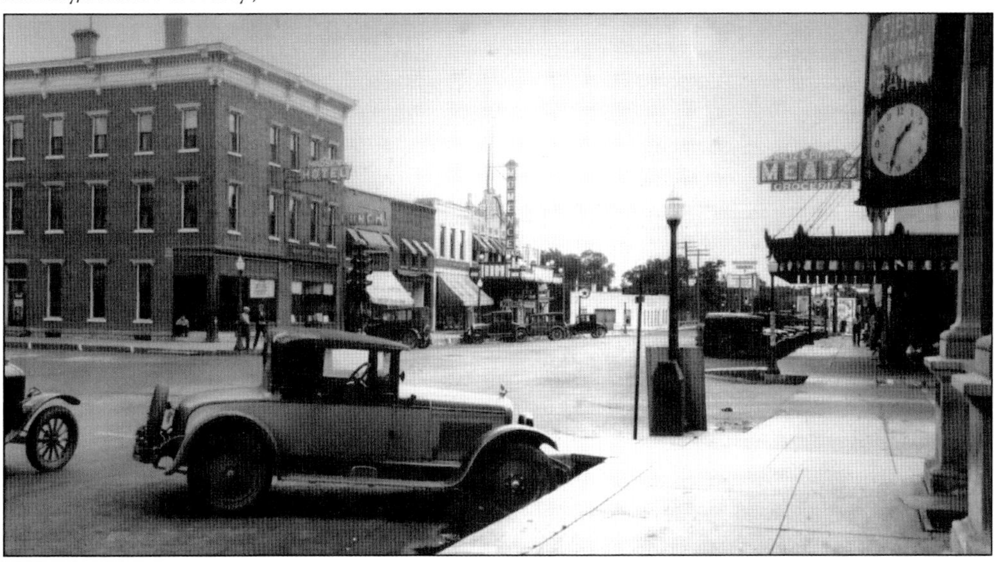

The building on the southeast corner of the Dixie Highway and Washington Street, built in 1882 by Fred Knighthart, has been a landmark in downtown Momence ever since. For years, the building was a hotel operating under names like the Central House, Madsen Hotel, Savoy Hotel, and Momence Hotel. Today a few retail shops occupy the first floor while apartments fill the rest of the building. (Courtesy Edward Chipman Public Library.)

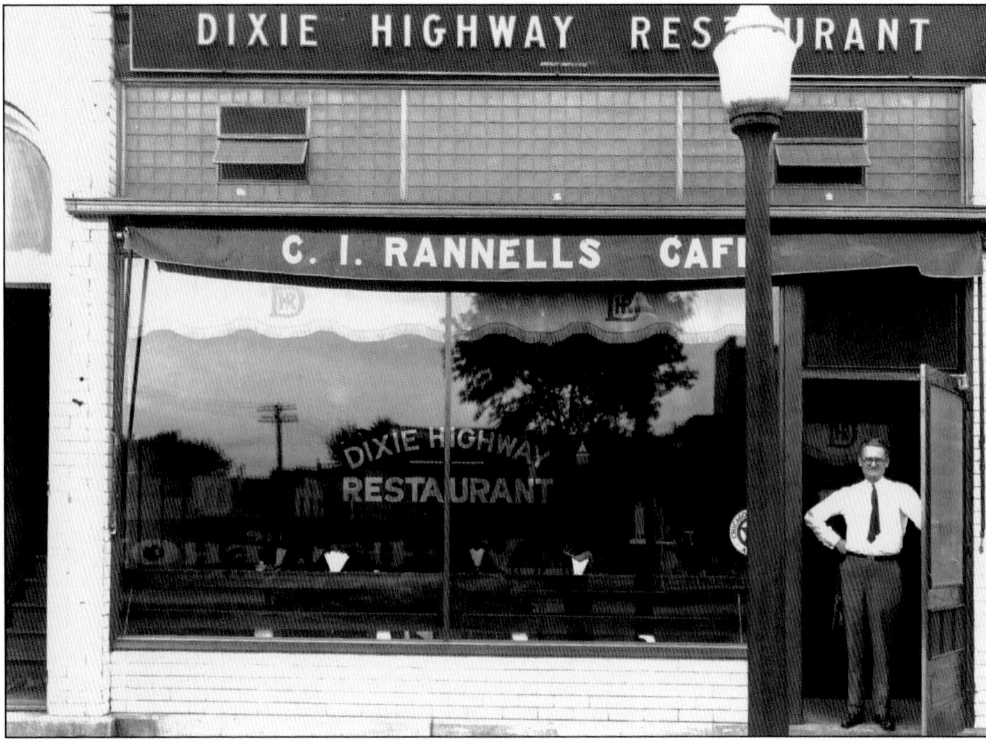

C. I. Rannells stands in front of the Dixie Highway Restaurant, a café he ran in Momence on the west side of the highway a few doors south of the Parish Bank. The Chicago Motor Club–approved restaurant offered travelers good food, clean facilities, and white tablecloth service at reasonable prices. The reflection of the photographer and his box camera can be seen in the window of this photograph from the late 1920s. (Courtesy Edward Chipman Public Library/Kevin McNulty.)

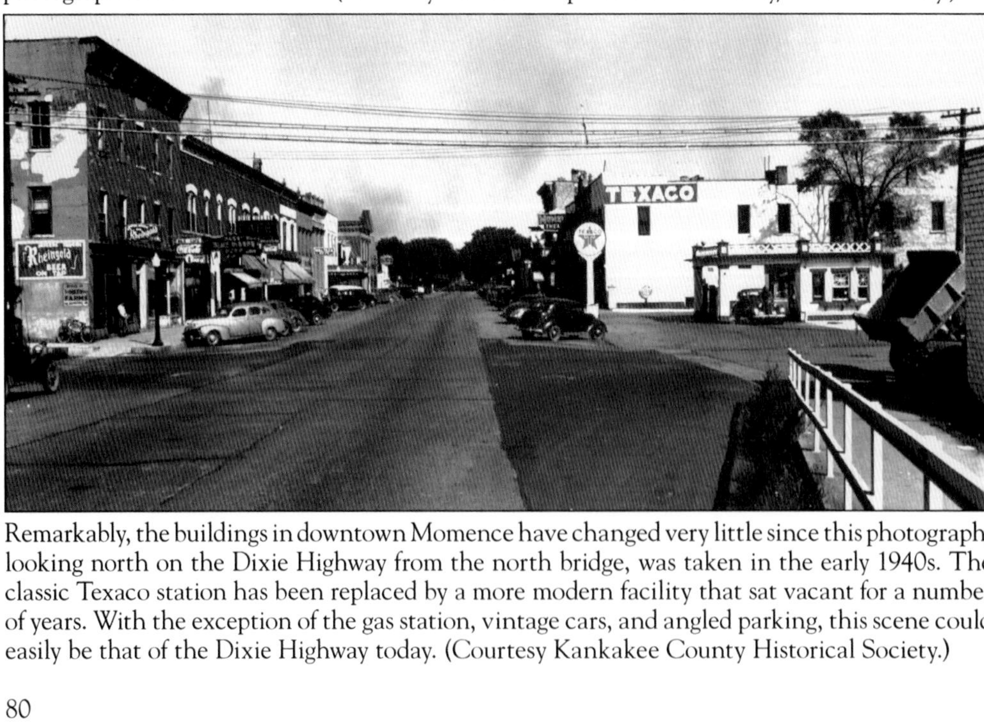

Remarkably, the buildings in downtown Momence have changed very little since this photograph, looking north on the Dixie Highway from the north bridge, was taken in the early 1940s. The classic Texaco station has been replaced by a more modern facility that sat vacant for a number of years. With the exception of the gas station, vintage cars, and angled parking, this scene could easily be that of the Dixie Highway today. (Courtesy Kankakee County Historical Society.)

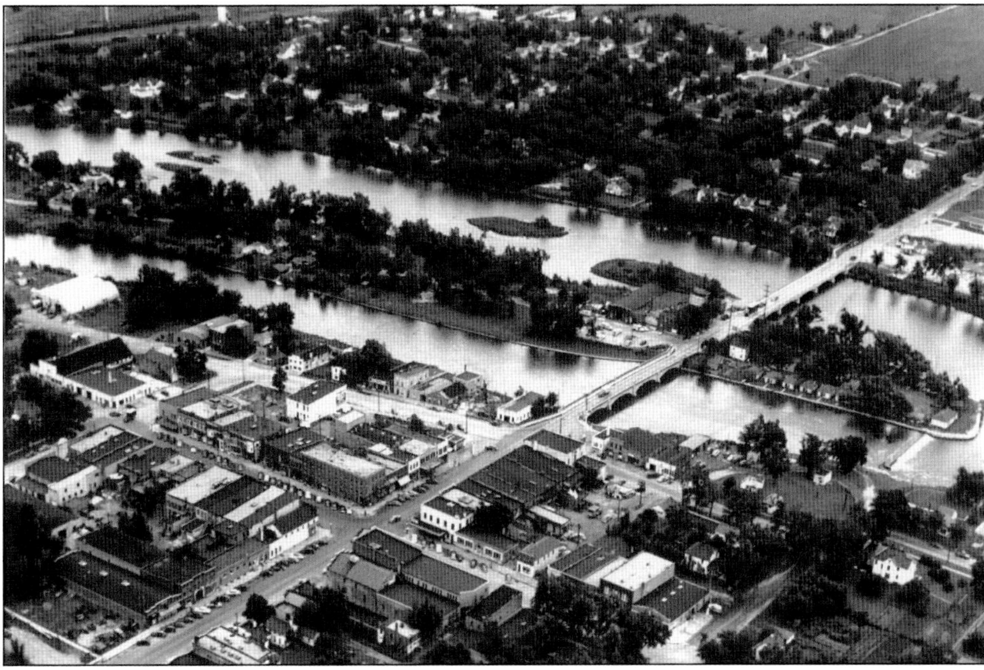

The Dixie Highway stretches across this view of downtown Momence taken after World War II. The photograph also provides a perfect image of the north and south bridges spanning the Kankakee River and the island located between them. Since French fur trade days, Momence has always been an important crossing on the river, which is relatively shallow at this point. (Courtesy Sue Butterfield/Kevin McNulty.)

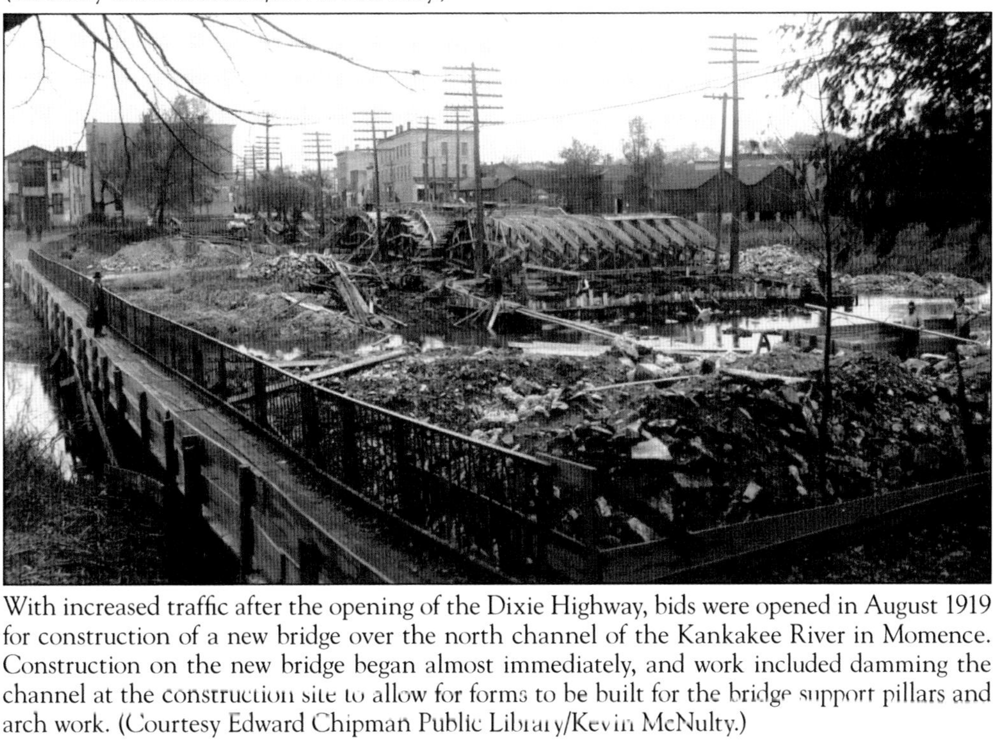

With increased traffic after the opening of the Dixie Highway, bids were opened in August 1919 for construction of a new bridge over the north channel of the Kankakee River in Momence. Construction on the new bridge began almost immediately, and work included damming the channel at the construction site to allow for forms to be built for the bridge support pillars and arch work. (Courtesy Edward Chipman Public Library/Kevin McNulty.)

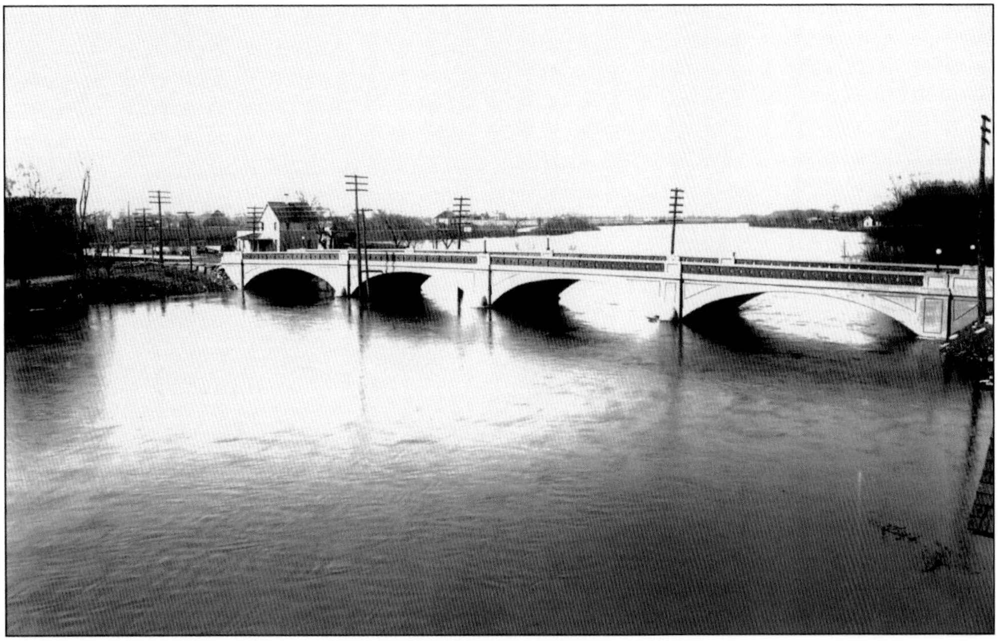

The bridge over the north channel of the Kankakee River in Momence was completed in 1920 by the M. P. Company at a cost of $23,789. This bridge replaced an old iron bridge with a wood deck that was built in 1872. The new bridge served well until 1985 when the current span over the north channel was built. (Courtesy Edward Chipman Public Library/Kevin McNulty.)

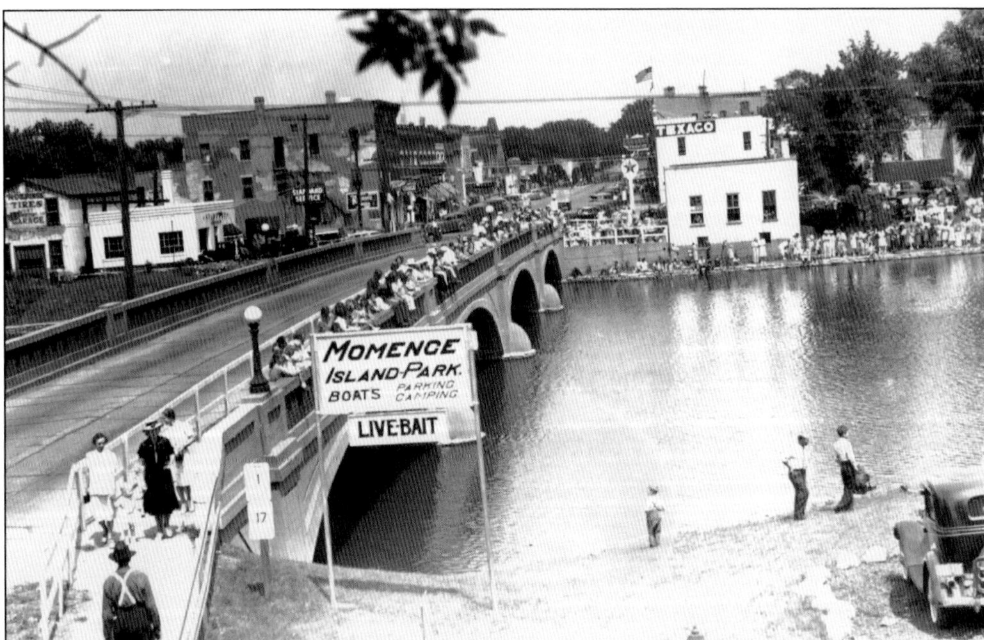

The Kankakee River has played an important role in the lives of Momence residents. Here spectators likely await the arrival of a river float during an early Gladiolus Festival. This photograph provides a wonderful summer view of the river, north bridge, Dixie Highway and downtown Momence. (Courtesy Edward Chipman Public Library/Kevin McNulty.)

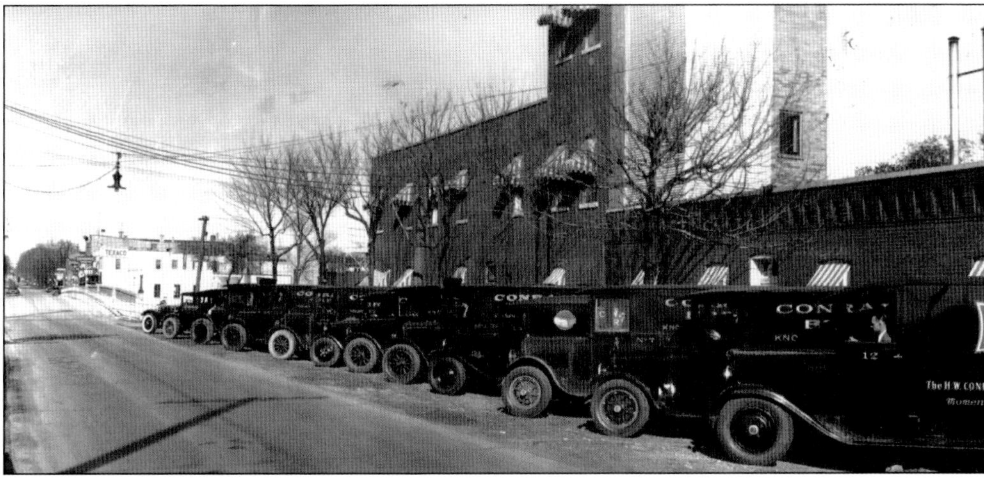

Oscar Conrad started his bakery business in Momence on the island between the north and south bridges in 1908. The bakery was the largest employer in town at one time. Henry Conrad later managed the business, which was sold to the Litoborski family after Henry's death in 1933. The building has since been torn down and the site, on the east side of the Dixie Highway at Mill Street, is now Conrad Park. (Courtesy Edward Chipman Public Library/Kevin McNulty.)

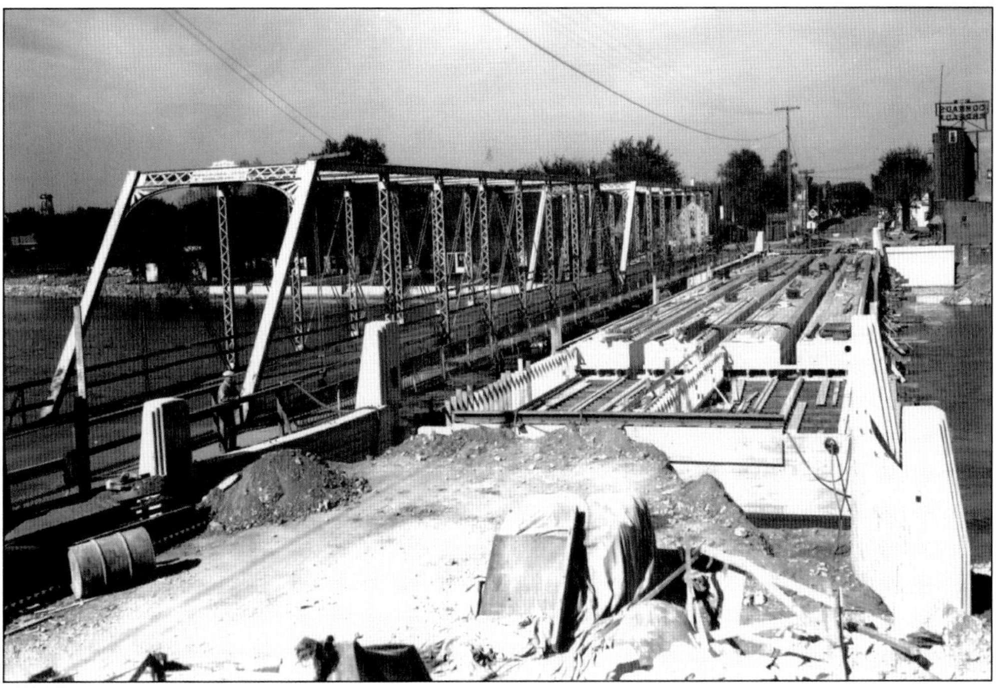

For decades, the Dixie Highway crossed the south channel of the Kankakee River in Momence on the two-span steel skeleton bridge (left) built by the Massillon Bridge Company in 1886. Prior to construction of the new concrete bridge (right) in 1938, the old bridge was moved slightly west and continued to serve as the temporary crossing for the Dixie Highway until the new bridge was completed. (Courtesy Kevin McNulty.)

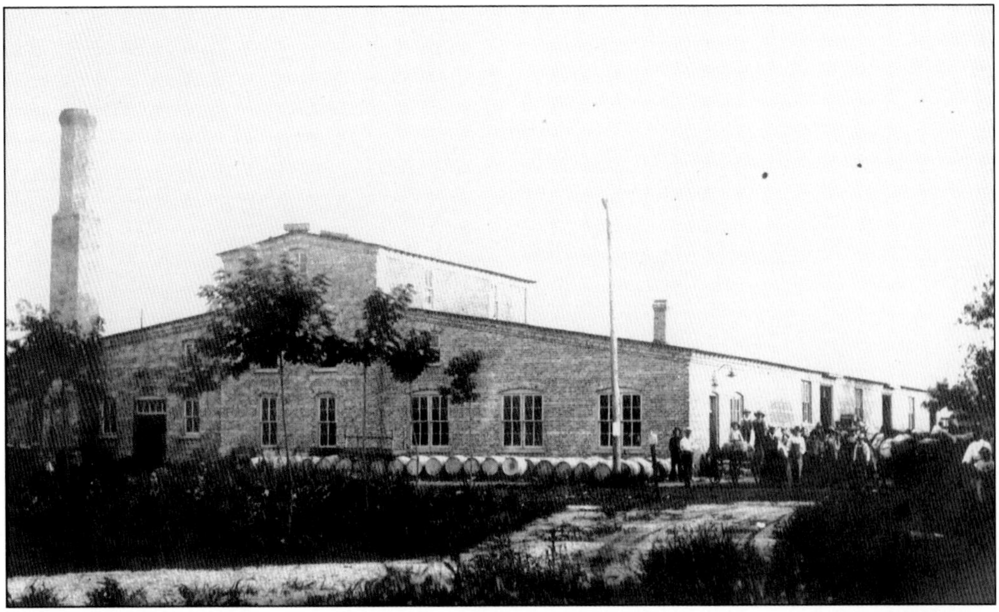

The original route of the Dixie Highway passed through Wichert, a tiny community in Kankakee County noted for growing a variety of vegetables. The original Claussen Pickle factory was located in Wichert. Built in 1905, this plant could process 25,000 bushels of pickles in season. The area was also known for producing asparagus, tomatoes, cabbage, and onion sets. (Courtesy Don Bonvallet.)

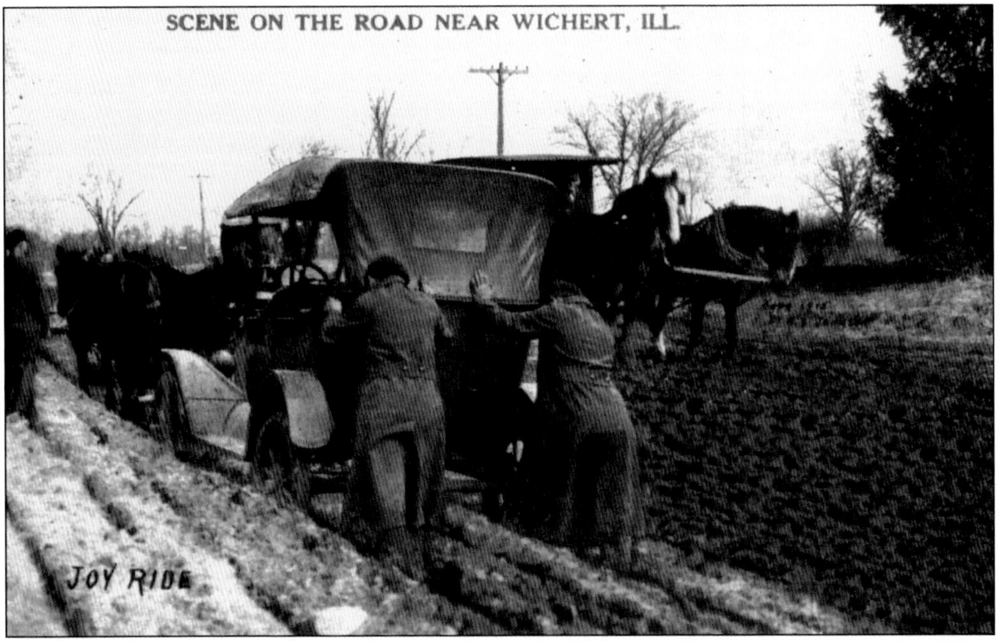

Before roads were paved, traveling by car or buggy could be an adventure in wet weather. Here two autoists get help from a local farmer pulling their vehicle out of the mud near Wichert. Although road conditions like this were not good, the Dixie Highway was rerouted to the west in 1919 to eliminate several grade rail crossings. (Courtesy Don Bonvallet.)

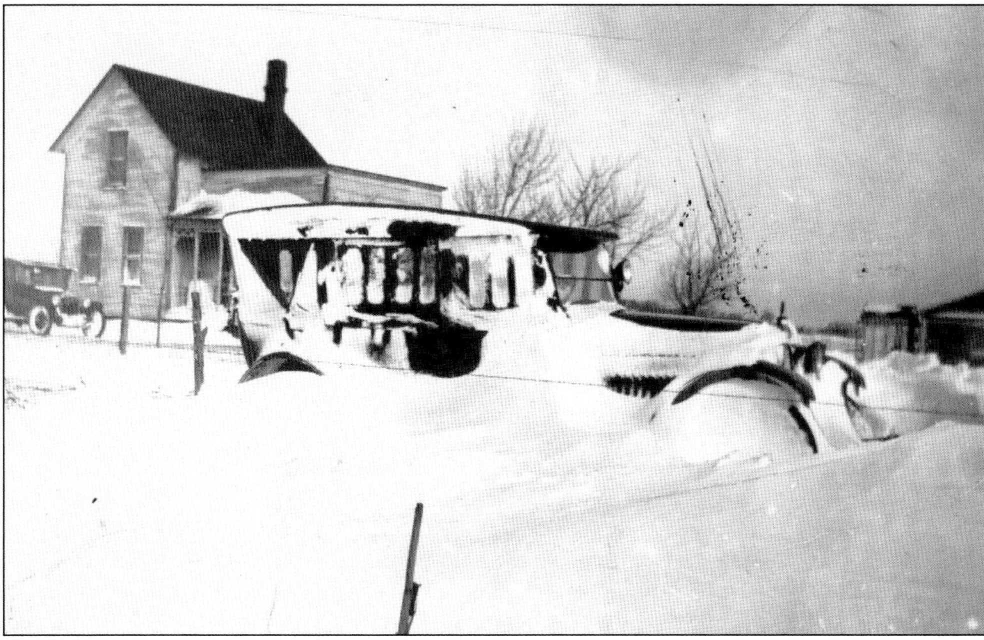

Wet road conditions were not the only hazard encountered by early motorists. With primitive or no snowplowing equipment at all, highway authorities were hard-pressed in their efforts to clear local roads. Winter driving frequently resulted in vehicles being stranded like this one near Wichert, especially on open stretches of rural roads. (Courtesy Don Bonvallet.)

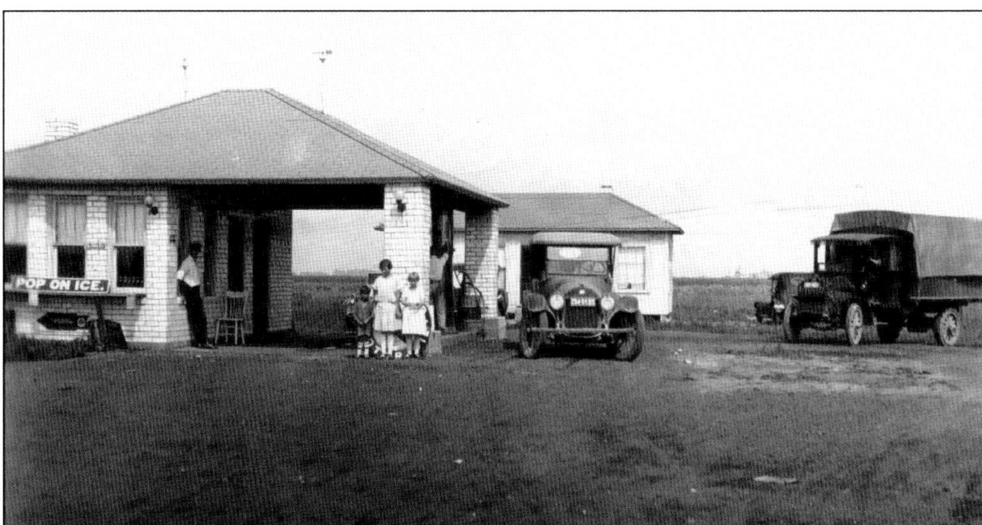

The intersection of Illinois Routes 1 and 17 in Kankakee County has changed greatly over the years. Known locally as the "Y" for its configuration, the junction has had a number of businesses catering to travelers. Fred and Georgia Langman built this gas station on the east side of the highway at the intersection. The Langmans' children, Bob, Loree, and Doris, are pictured from left to right. For a while, the family lived in the small house behind the station. (Courtesy Brent Langman family.)

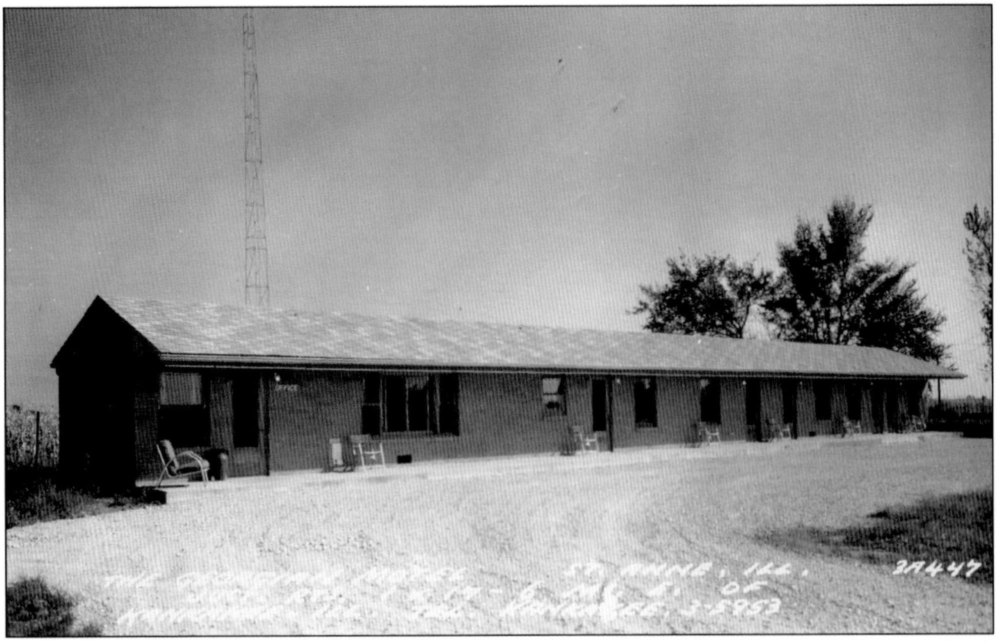

After her husband's death, Georgia Langman leased her gas station to another operator and built this modern motel at the Y north of St. Anne in 1953. Langman ran the motel for a number of years as the Georgian and later sold the property. The motel is no longer in business, but the building still stands, now south of the current configuration of the Illinois Routes 1 and 17 intersection. (Courtesy Brent Langman family.)

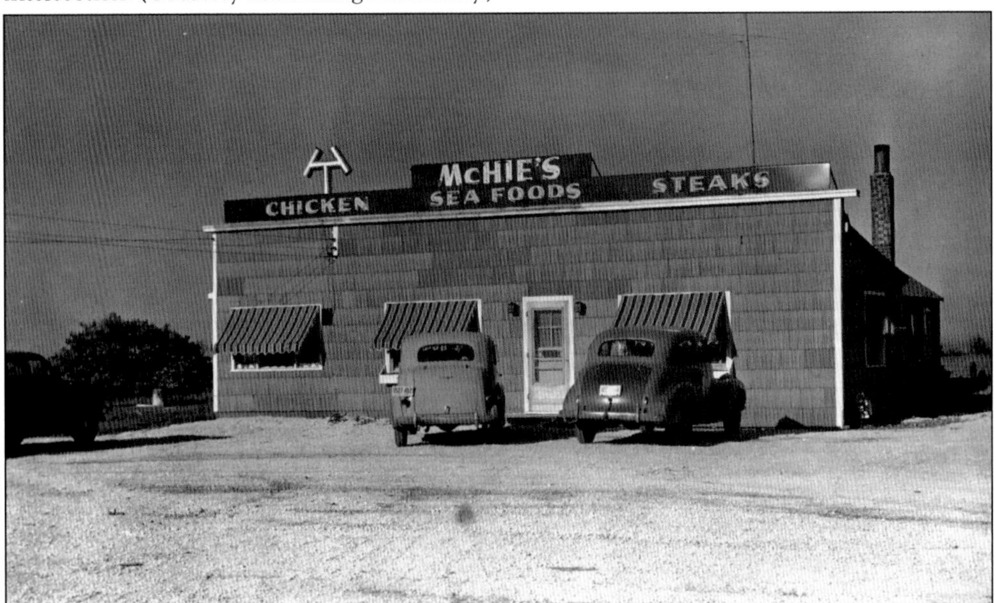

South of the Langman gas station on the east side of Illinois Route 1, the Dixie Highway, at the Illinois Route 17 intersection, Sidmon and Mina McHie ran this restaurant and tavern with a dance floor catering to travelers and local clientele alike. The restaurant north of St. Anne was well known for its plate lunches and steak, chicken, and seafood dinners, all served at reasonable prices. The McHies ran the restaurant until 1954. (Courtesy Richard McHie.)

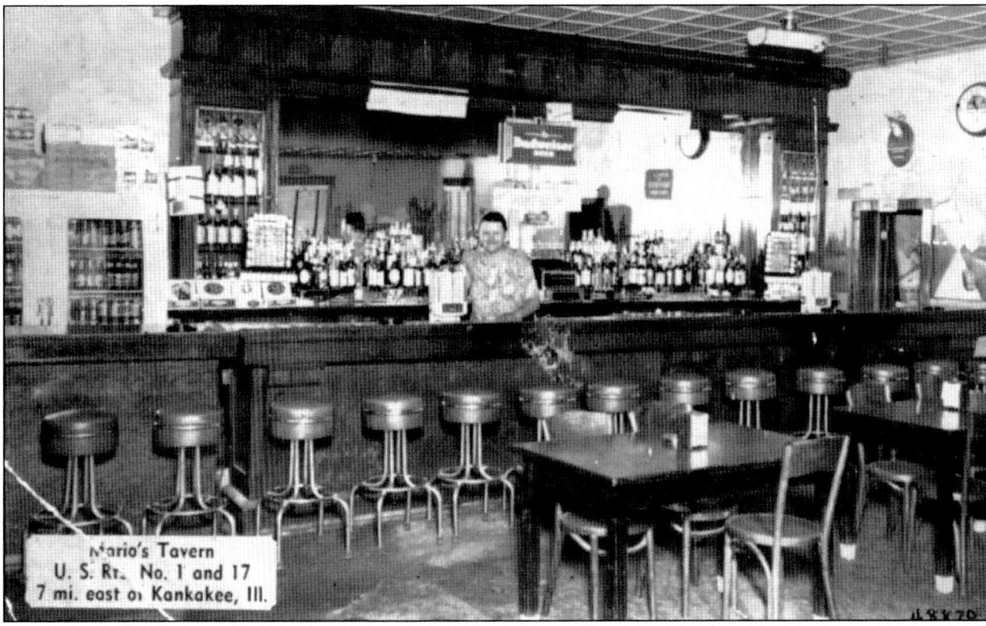

Perhaps the most well-known business at the Y, the intersection of Illinois Routes 1 and 17, was Mario's Restaurant and Tavern. Owned by Mario Viglielmo and his wife, Delores, since 1946, Mario's was open 24 hours a day, 365 days a year and was a rest stop for long-distance buses and truckers. After interstate highways siphoned traffic off Illinois Route 1, business declined and Mario's closed in the mid-1980s. The building has since been demolished. (Courtesy Robert Hoekstra.)

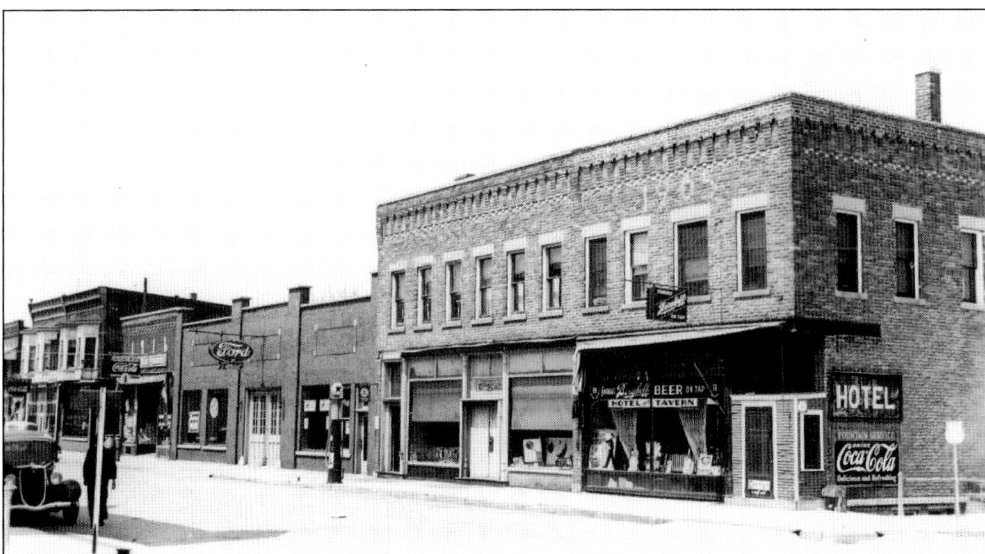

From 1915 through 1920, the original route of the Dixie Highway passed through Wichert and proceeded south on what is now 8500E Road then west on 6750S Road, joining Station Street through downtown St. Anne. The road then followed the current alignment of the highway to Guertin Street and continued south on 7500E and Cable Line Roads into Papineau. The north side of Station Street looking west in downtown St. Anne is pictured. (Courtesy Robert Bovie.)

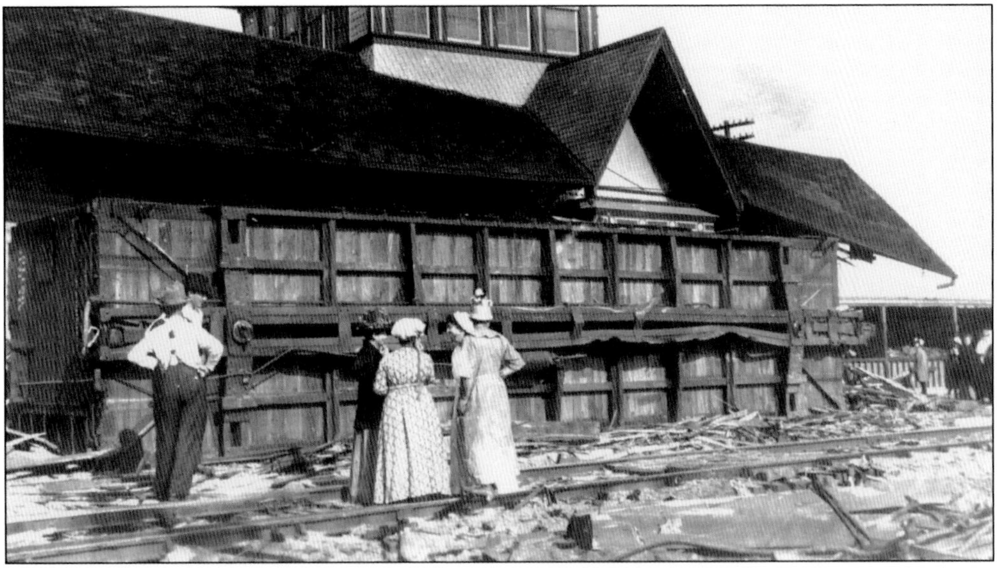

Government engineers objected to the original route of the Dixie Highway between St. Anne and Momence. The route, which passed through Wichert, crossed the C&EI and the Cleveland, Cincinnati, Chicago and St. Louis Railway (Big Four) tracks six times. The current alignment of the highway, with only one rail crossing, was selected in 1918 but not fully paved until the end of 1920. Trains were not only dangerous to motorists; in July 1917, a freight train derailed just north of the Dixie Highway crossing and severely damaged the St. Anne depot. (Courtesy Don Bonvallet.)

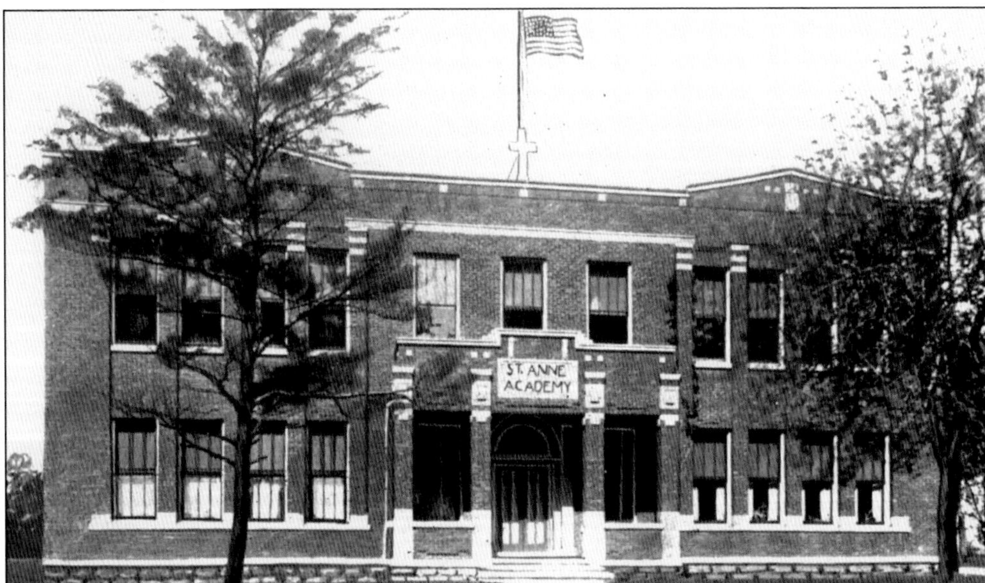

Early settlers in the St. Anne area were Roman Catholics from Canada. They built a church in honor of St. Anne, and the growing community became known by that name also. Church members established a school in 1884. The original building burned down in November 1920. This academy building was completed in September 1921 and was located just east of the Dixie Highway and north of Grant Street. The school closed in 1978, and the building has since been torn down. (Courtesy Robert Bovie.)

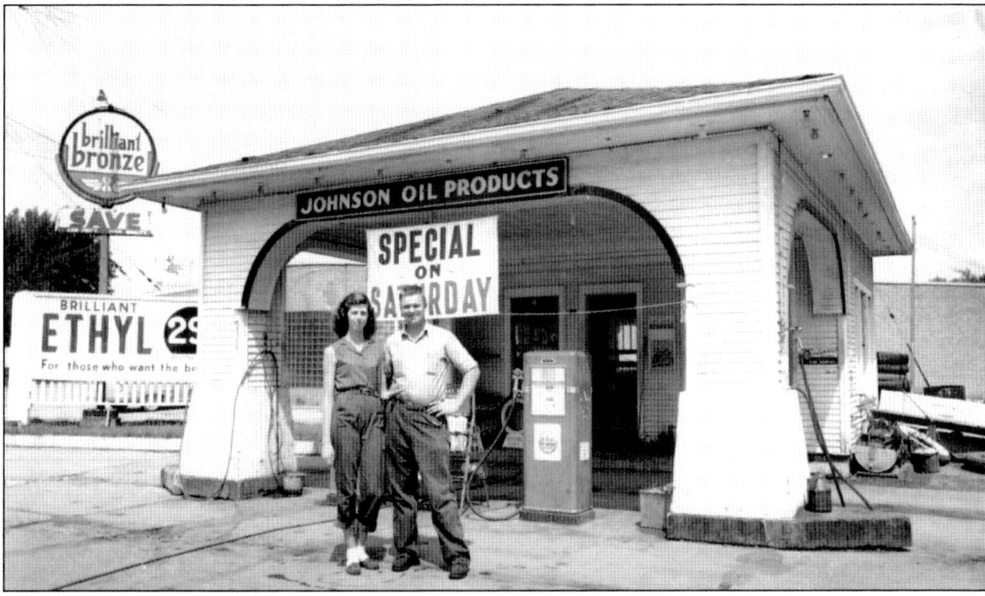

Deloris and Glen Redenius pose in front of their gas station in St. Anne on the northeast corner of the Dixie Highway, Illinois Route 1, and Station Street. Johnson Oil, a refiner that marketed throughout the Midwest, began operations in 1909. Its original bulk plant was in South Chicago Heights. Company headquarters later moved to Chicago Heights and finally to Chicago. The company marketed Brilliant Bronze and Ethyl gasolines and ceased operation in 1956. (Courtesy Doug Redenius.)

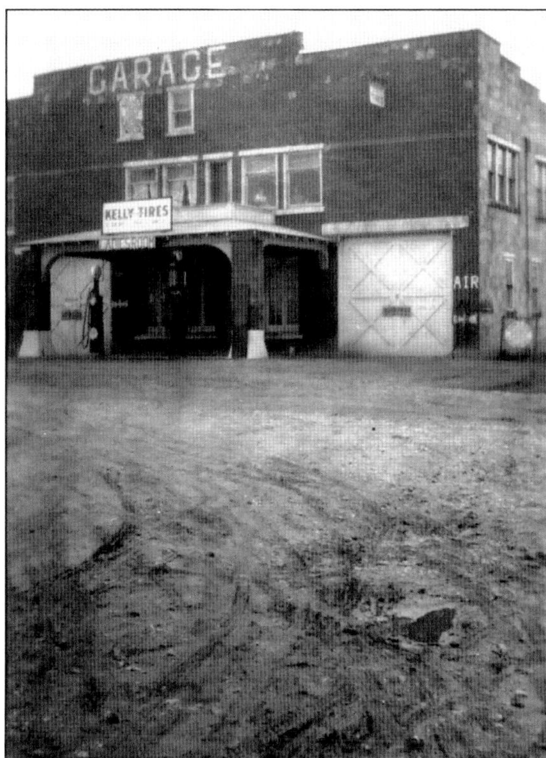

Brothers Carl and Henry Allain owned this garage in St. Anne on the northeast corner of Illinois Route 1, the Dixie Highway, and Guertin Street. The brothers did repairs, sold Sinclair brand gas and oil products, and were dealers for Kelly tires. The building, a landmark at the intersection, burned down in December 1952, and a Shell gas station and convenience store now occupies the site. Brothers Dale and Gilbert Irps can be seen peering from the second-floor windows. (Courtesy Gilbert Irps.)

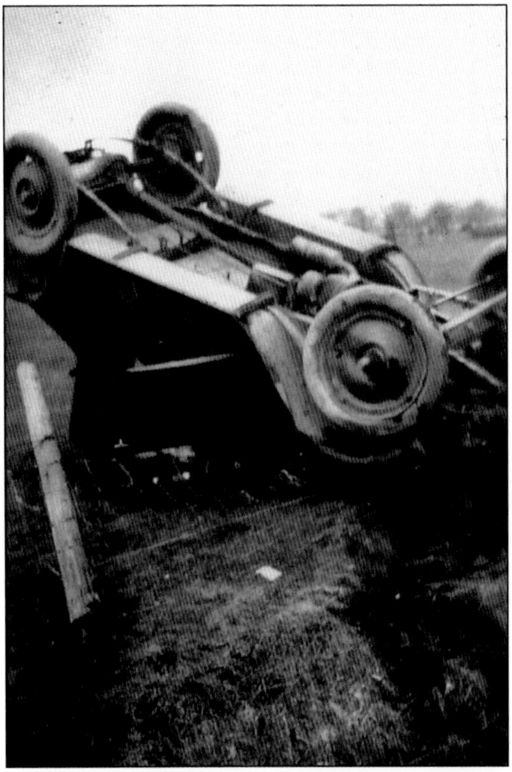

The curve on the Dixie Highway, Illinois Route 1, heading west at Guertin Street on the south end of St. Anne was always treacherous until the intersection was widened in the 1950s. It was not uncommon for southbound vehicles traveling at night or going too fast for conditions to miss the curve and "turn turtle" like this one and end up in a farmer's pasture. (Courtesy Gilbert Irps.)

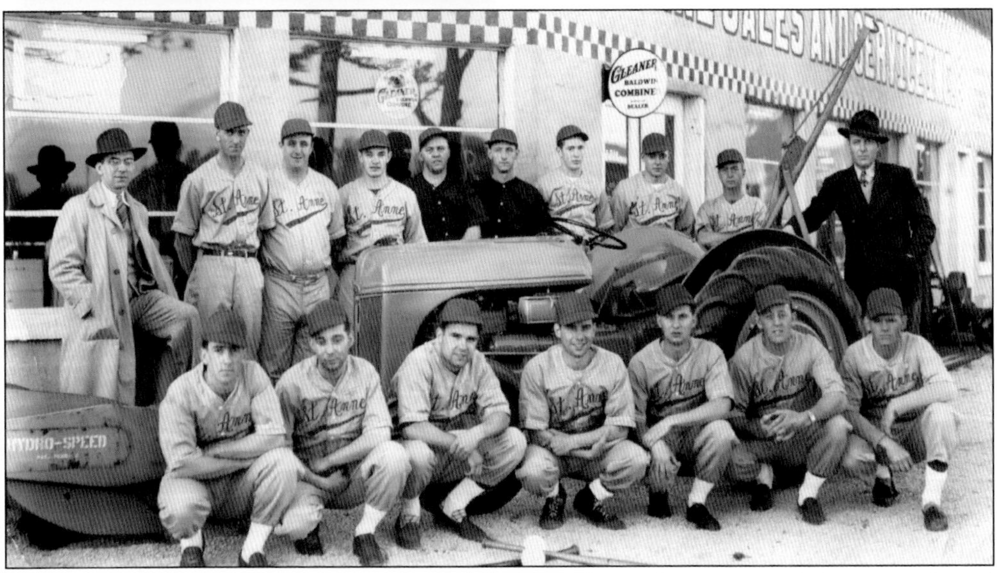

St. Anne Sales and Service, dealers in Ford cars, trucks, and tractors and Purina seeds and feed, was located on the northwest corner of the Dixie Highway curve on the south end of town. The company sponsored a baseball team for years. Pictured here are, from left to right, (first row) Kenneth Tunks, "Pickles" Olena, George Weber, Dick Weber, Stanley DeWitt, Lester Dykstra, and Clarence Cadwaller; (second row) L. D. Corkins, Al Slingerland, "Moose" Manny, Jerry Bovie, "Mits" Mercy, Leo Hebert, "Red" Ahrens, Bob Easter, Gil Hanen, and Ike Parrish. (Courtesy Kenneth Tunks.)

Four

Iroquois County

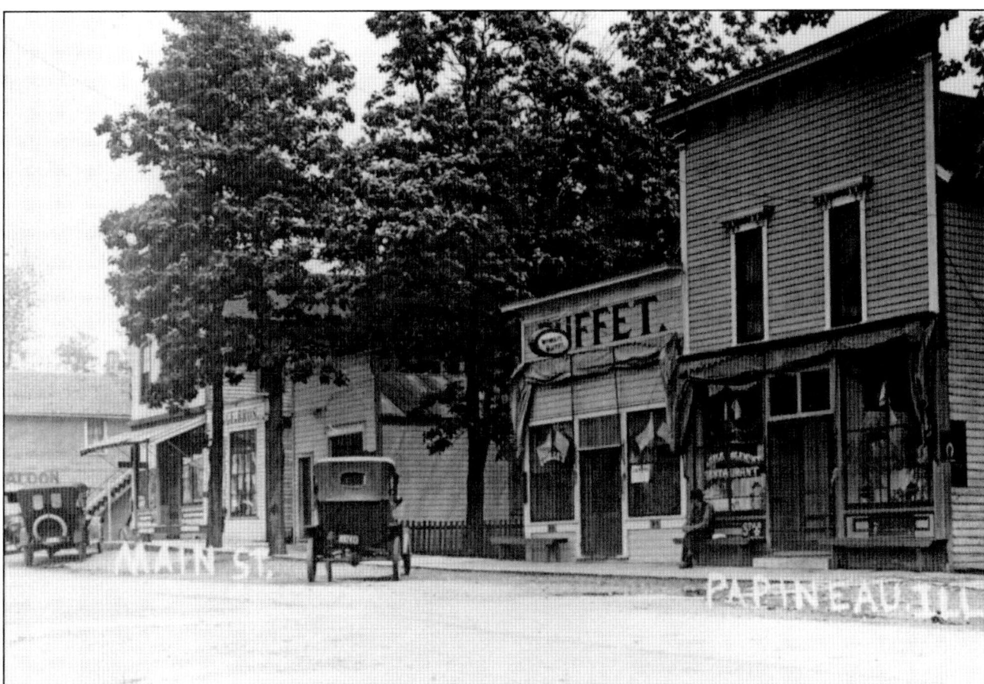

South of the Kankakee-Iroquois County line, the original alignment of the Dixie Highway followed Main Street, now East Avenue, through Papineau from 1915 to 1920. After concrete paving was completed on the current alignment in 1920, downtown Papineau was bypassed. Papineau had a number of businesses catering to passing motorists including the William Wolfe Buffet and Jule Clement's Restaurant. (Courtesy Marilyn Miller.)

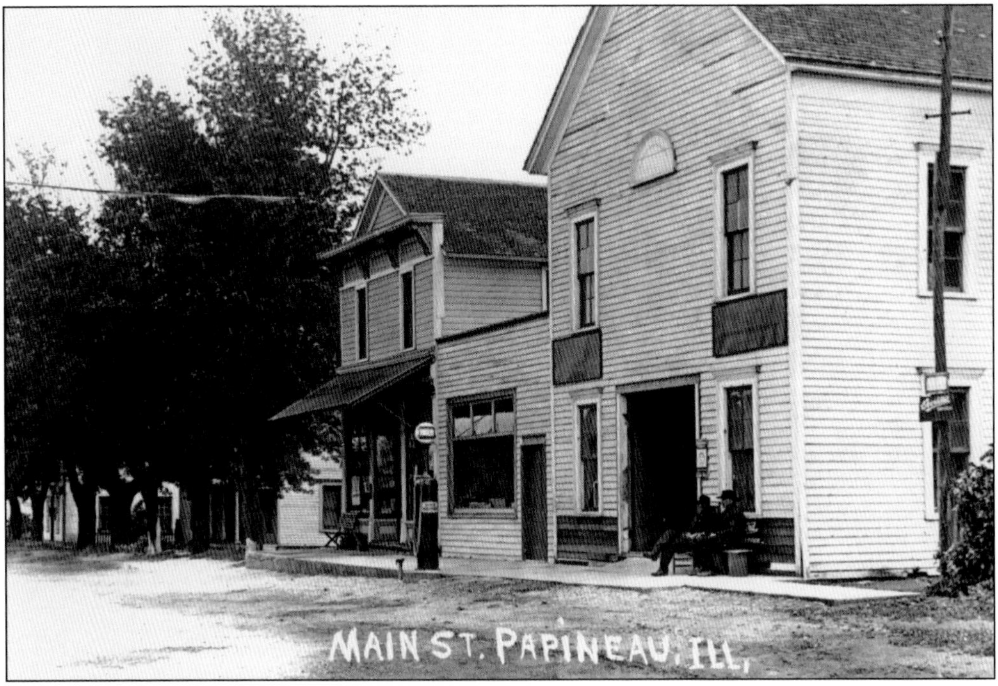

A couple of men relax in front of the Papineau Garage on Main Street (now East Avenue). Although hard to see in the photograph, the gas pump in front of the garage sports the original cotton bale logo of the Dixie Highway. Dixie Highway traffic traveled through Papineau until 1920, when the current road alignment was completed. (Courtesy Marilyn Miller.)

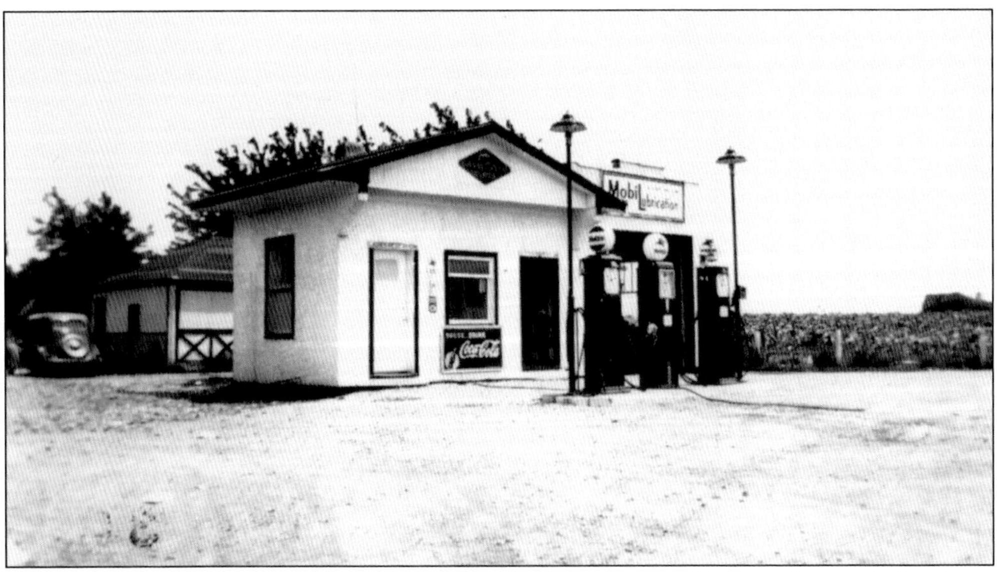

This small Mobil service station was located on the east side of Illinois Route 1, the Dixie Highway, at Main Street in Martinton. The station was located just south of the Cozy Inn, which later became Johnnie's Trading Post. Although small, the station even had a ladies' restroom, a clean and welcome amenity for the traveling public. (Courtesy Jeff Webster.)

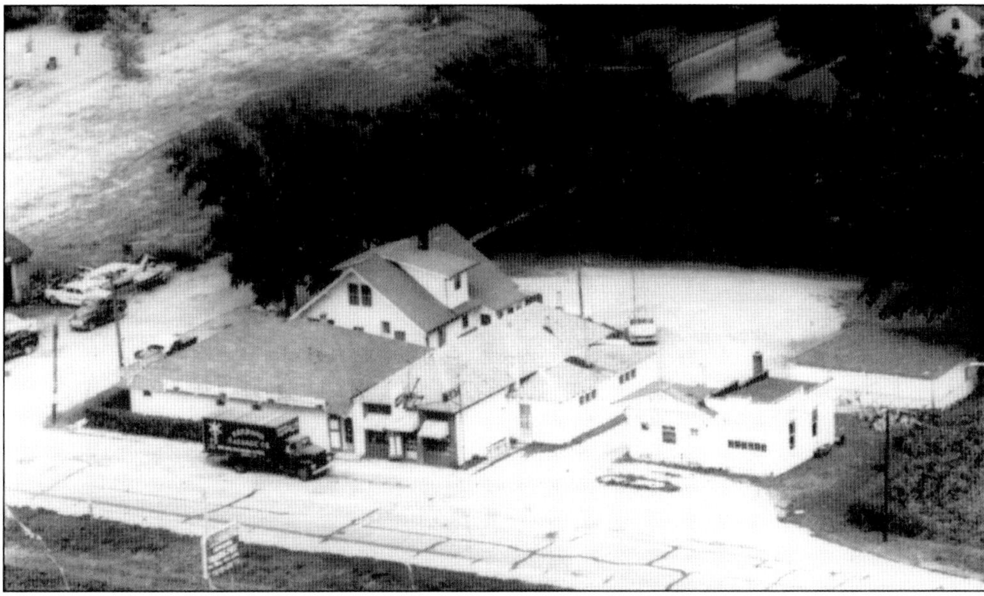

Johnnie's Trading Post is pictured in this aerial view in Martinton. Owned by Johnnie Newman, the business originally was known as the Cozy Inn and consisted of a restaurant, bar, and dance hall. In the 1950s, Newman ceased operating the restaurant and dance hall and converted the space to his "trading post," selling a variety of necessities and gift items. The buildings were torn down in 1997. (Courtesy Jeff Webster.)

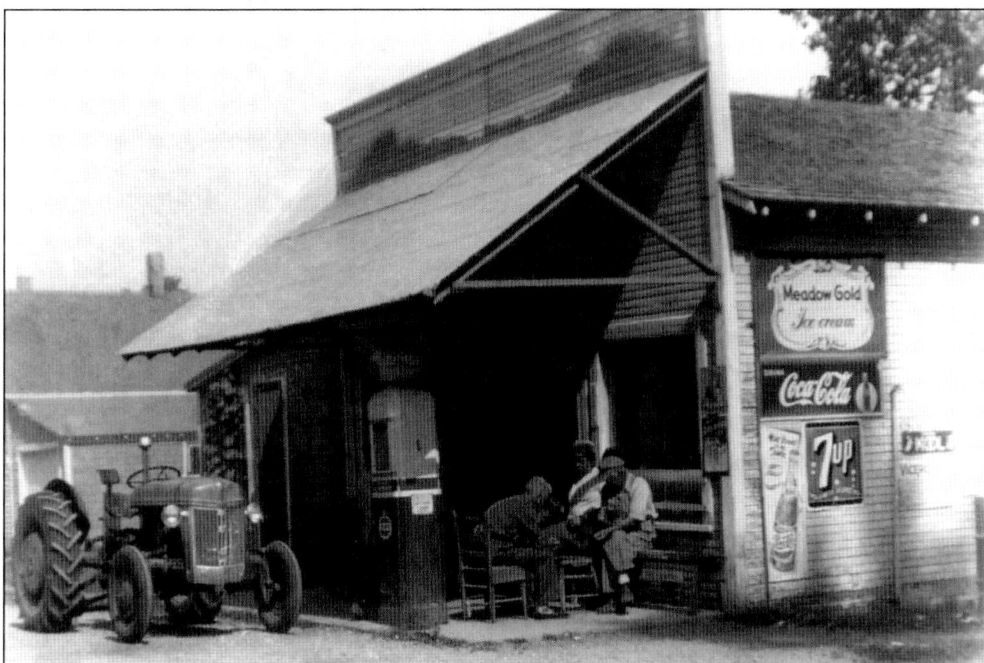

John Peters and Harry Creps play a game of checkers in front of Peters' General Store east of Illinois Route 1, the Dixie Highway, in Pittwood. The highway made two 90-degree turns just west of this small crossroads located south of U.S. Route 52 in Iroquois County. The dangerous turns were smoothed out with a sweeping curve in the mid-1950s. (Courtesy Jim Pence.)

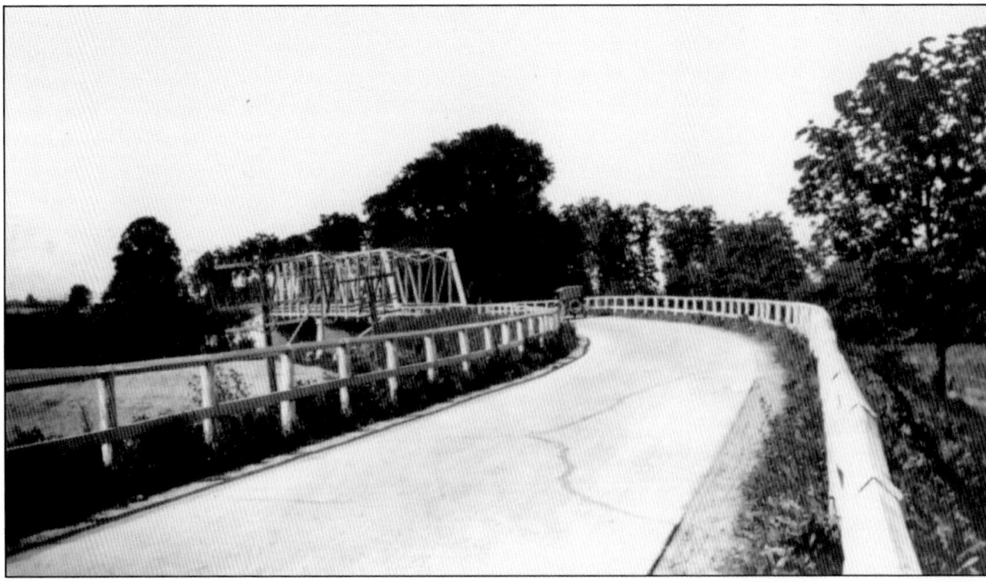

A lone automobile travels northbound on Illinois Route 1, the Dixie Highway, approaching the bridge over the Iroquois River, north of Watseka. The bridge was completed in April 1921. A new bridge was built just west of this location in 1957. The 1921 bridge has been demolished; however, the bridge piers can still be seen today at the river's edge east of the roadway. (Courtesy Jeff Webster.)

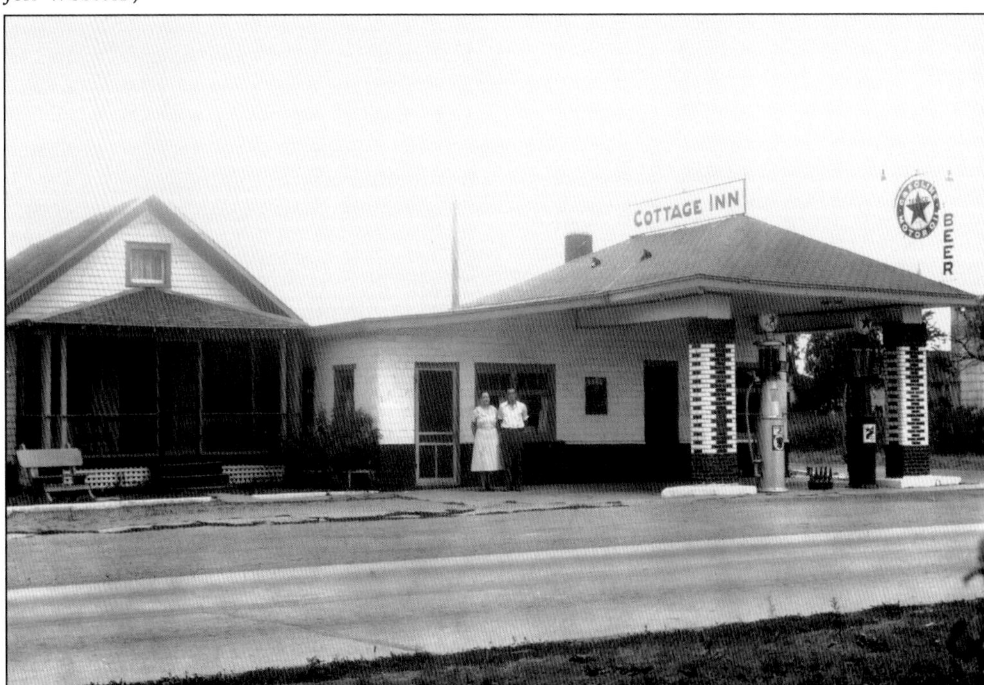

Mildred and Bob Cavitt stand in front of the Cottage Inn diner and Texaco service station they operated on Illinois Route 1, the Dixie Highway, north of Watseka. Like many early gas stations, the family lived next door. Texaco Fire Chief gasoline was 16¢ per gallon and Texaco Ethyl was 18¢ per gallon when this photograph was taken in the 1930s. (Courtesy Jim Pence.)

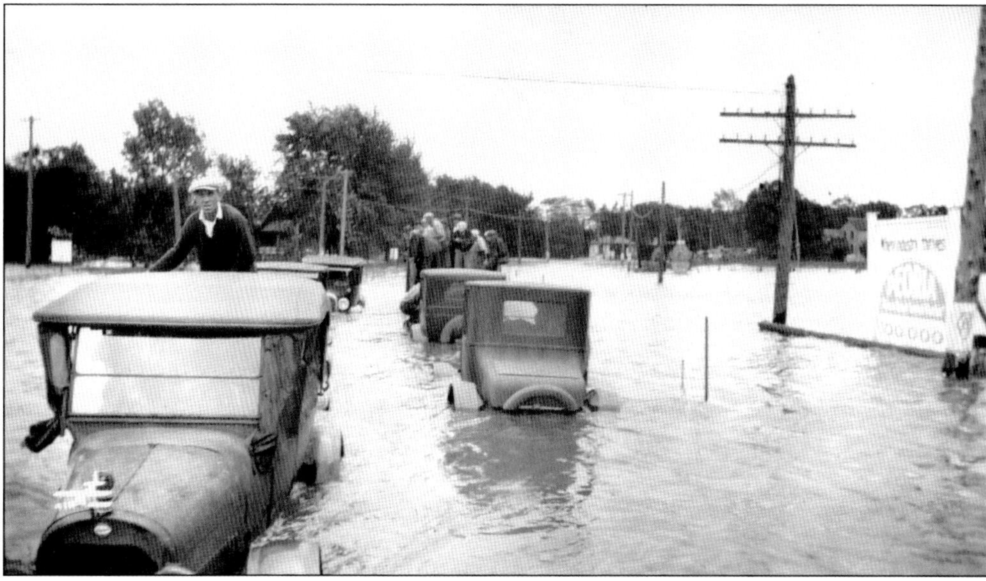

Flooding has always been a problem in Watseka, particularly near the intersection of Illinois Route 1 and U.S. Route 24. In this October 1926 view, courageous motorists attempt to ford the floodwaters and travel west through the intersection. At this time, U.S. Route 24 was known as Illinois Route 8. (Courtesy Iroquois County Genealogical and Historical Societies.)

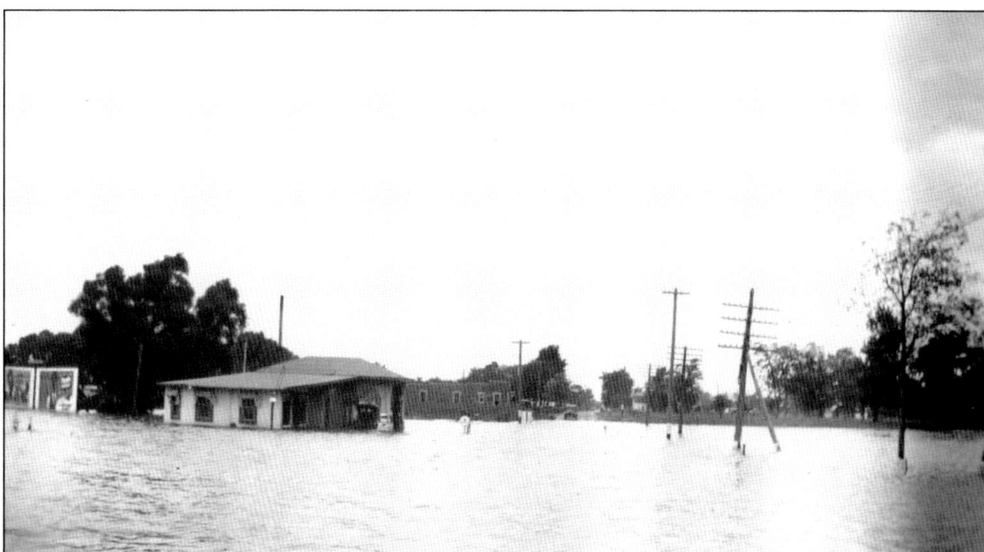

Floodwaters prevent a motorist from traveling any farther south on Illinois Route 1, the Dixie Highway, just north of Illinois Route 8, now U.S. Route 24. O'Neill's Service Station and Restaurant, a popular stop for years in Watseka at the northwest corner of the intersection, is at left. (Courtesy Iroquois County Genealogical and Historical Societies.)

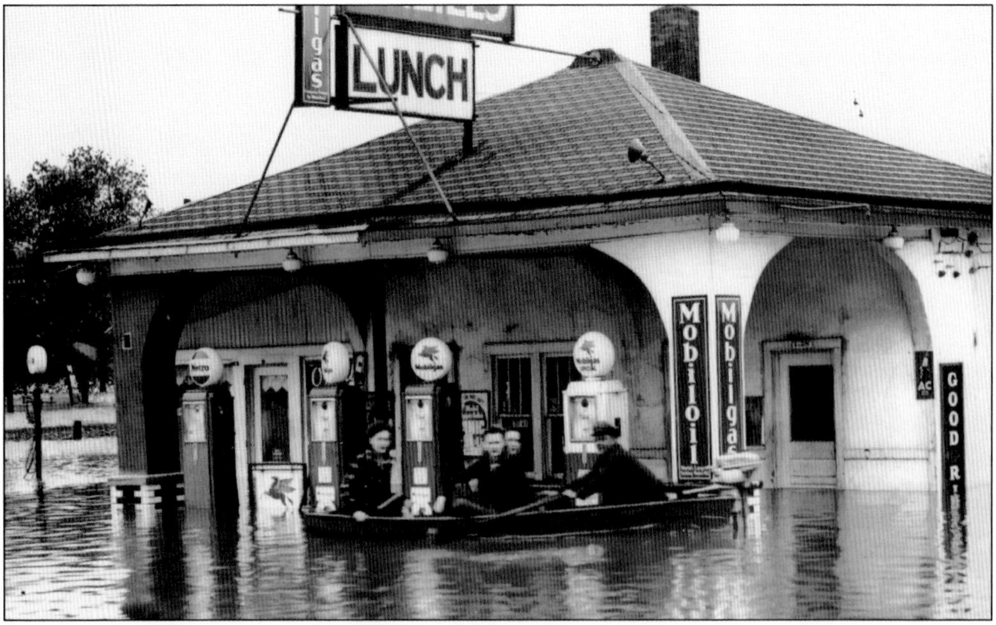

O'Neill's Service Station always took a beating when the Iroquois River overflowed its banks in Watseka. Here owner William O'Neill mans the oars on a rowboat in front of his station during a flood in the 1950s. The Watseka Family Table Restaurant occupies higher ground north of where the station once stood. (Courtesy Watseka Family Table Restaurant.)

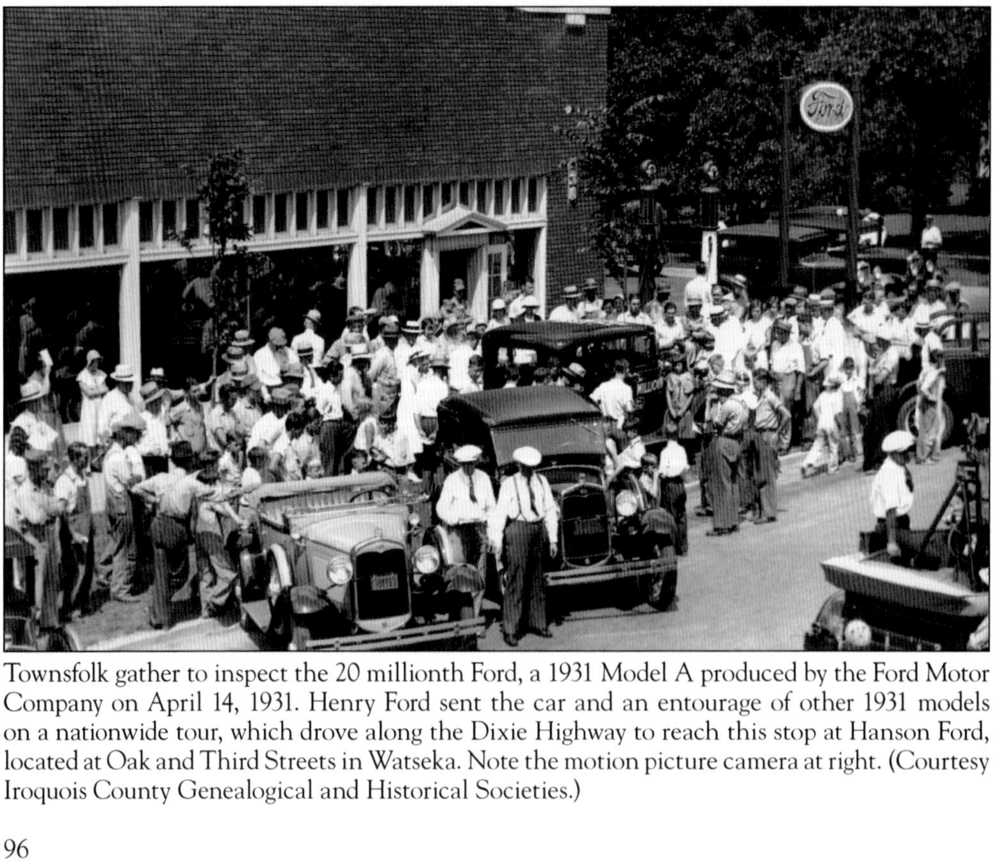

Townsfolk gather to inspect the 20 millionth Ford, a 1931 Model A produced by the Ford Motor Company on April 14, 1931. Henry Ford sent the car and an entourage of other 1931 models on a nationwide tour, which drove along the Dixie Highway to reach this stop at Hanson Ford, located at Oak and Third Streets in Watseka. Note the motion picture camera at right. (Courtesy Iroquois County Genealogical and Historical Societies.)

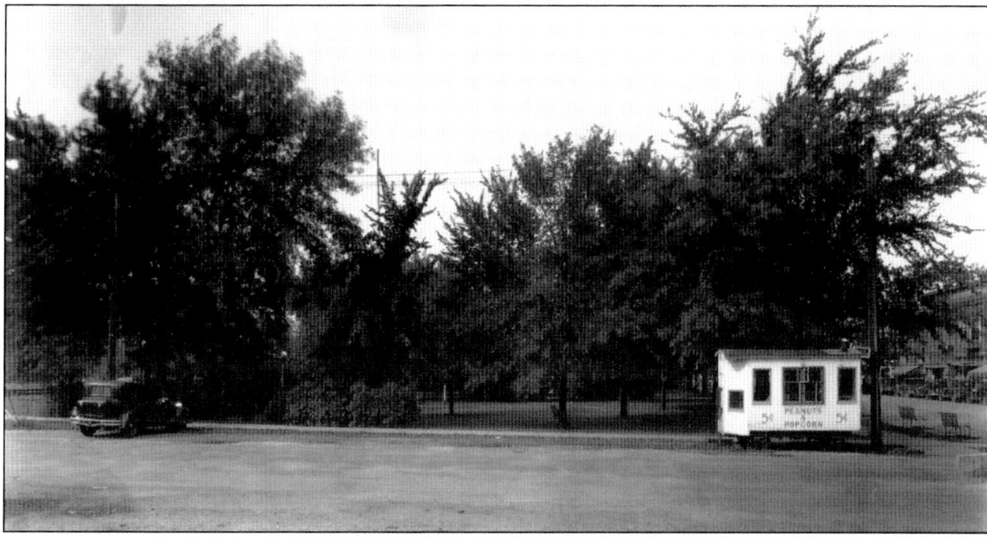

Woman's Club Park provided a peaceful respite for tourists and Watseka citizens alike. The park was located on the south side of Walnut Street, the Dixie Highway, between Second and Third Streets. The popcorn and peanut stand was a regular fixture in the park and is parked on the corner of Third and Walnut Streets in this view looking west on the Dixie Highway. (Courtesy Iroquois County Genealogical and Historical Societies.)

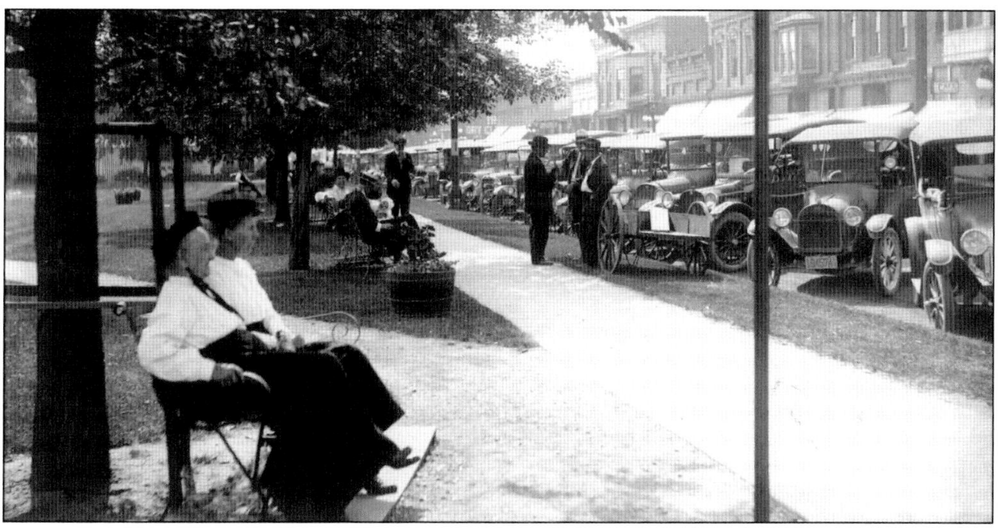

The south side of Walnut Street, the Dixie Highway, was jammed with cars parked along the northern boundary of Woman's Club Park in Watseka. Local residents and farmers were likely in town for a Saturday shopping trip or a holiday when this photograph was taken. Many seem to enjoy just "taking it all in" while sitting on a park bench under a shade tree. (Courtesy Iroquois County Genealogical and Historical Societies.)

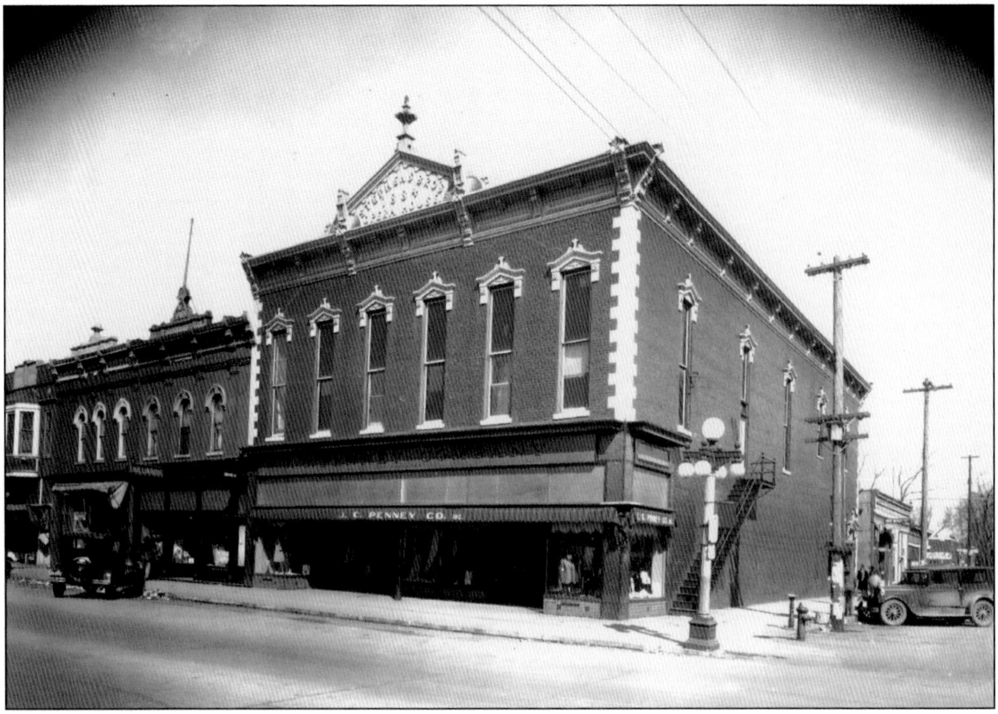

The J. C. Penney store was one of the many retail shops that lined Walnut Street, the Dixie Highway, in downtown Watseka. J. C. Penney's was located at the northwest corner of Third and Walnut Streets and occupied a building, built by the Stephens Brothers in 1884, that once housed a 700-seat opera house on the second floor. (Courtesy Iroquois County Genealogical and Historical Societies.)

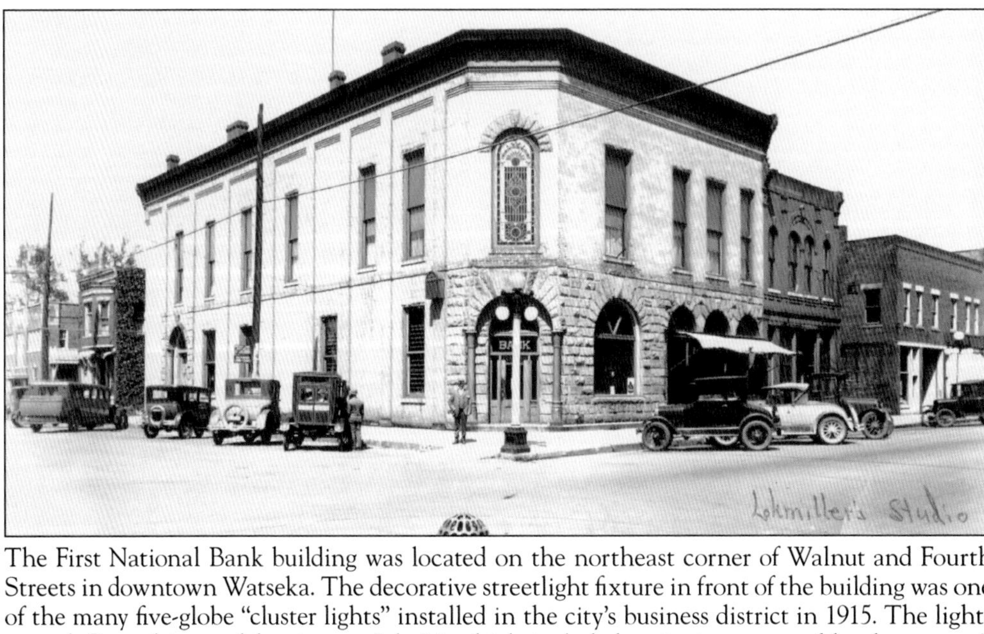

The First National Bank building was located on the northeast corner of Walnut and Fourth Streets in downtown Watseka. The decorative streetlight fixture in front of the building was one of the many five-globe "cluster lights" installed in the city's business district in 1915. The lights were dedicated in a celebration on July 14, which included a picnic sponsored by the county's Dixie Highway boosters. (Courtesy Iroquois County Genealogical and Historical Societies.)

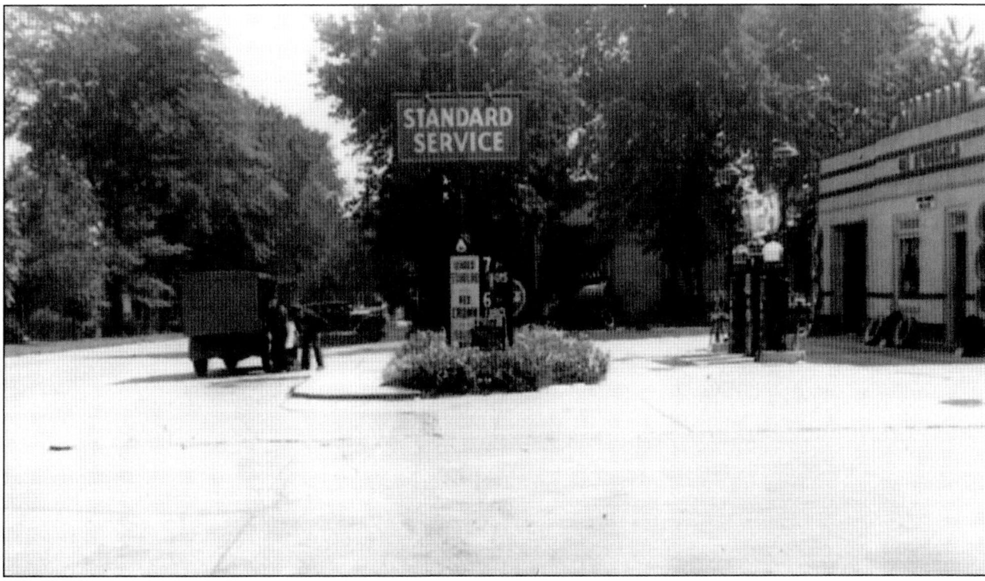

Vince O'Brien owned this modern Standard service station in Watseka at the southeast corner of Fifth and Walnut Streets. Atlas tires were a big seller at the station, which was selling seven gallons of leaded Standard for $1.05 when this photograph was taken in the 1930s. This view looks east down Walnut Street, the Dixie Highway. Jim Pence has owned the station since the 1960s. (Courtesy Jim Pence.)

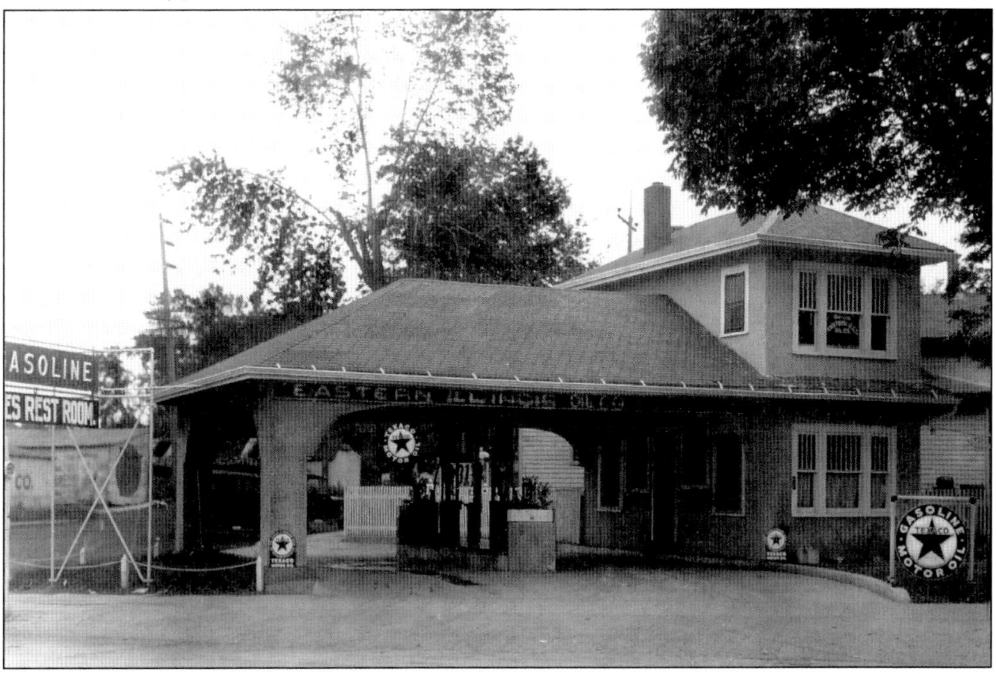

Gas stations were plentiful along the Dixie Highway in Watseka. Most major brands were represented, as was Texaco with this interesting station design located on Walnut Street. The two-story station with its large canopied drive was unusual for the time in that it allowed motorists to gas up sheltered from the elements, something that is taken for granted today. (Courtesy Jim Pence.)

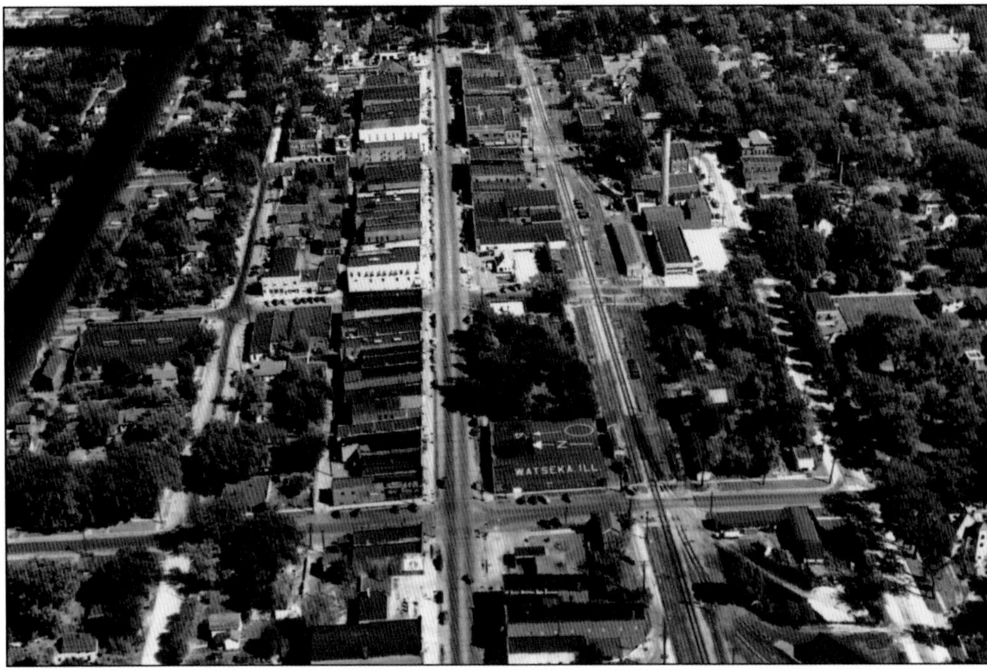

As the seat of Iroquois County, Watseka has always been the retail and commercial hub of the surrounding area. This aerial view looks eastward showing the bustling business district centered along Walnut Street, the Dixie Highway, in the center of the photograph. (Courtesy Iroquois County Genealogical and Historical Societies.)

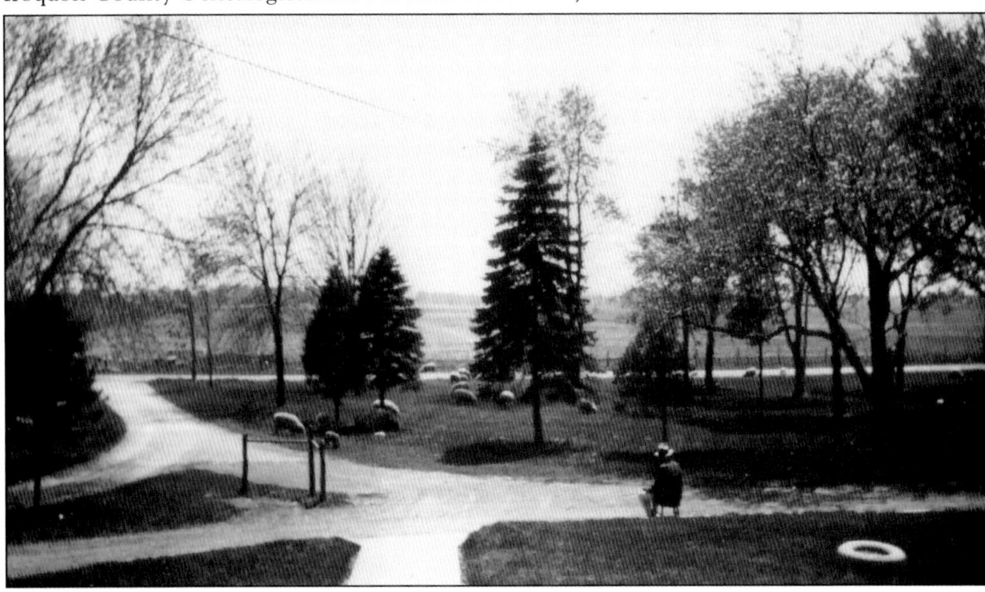

The original route of the Dixie Highway through Watseka traveled south on Fifth Street to Belmont Avenue then east on County Farm Road, passing south of the county home as shown here. On July 14, 1915, Speaker of the House and Danville resident "Uncle Joe" Cannon turned the first ceremonial spadeful of dirt to signal the start of construction of the highway in Illinois at the intersection of County Farm Road and what is now Illinois Route 1. (Courtesy Larry Waters.)

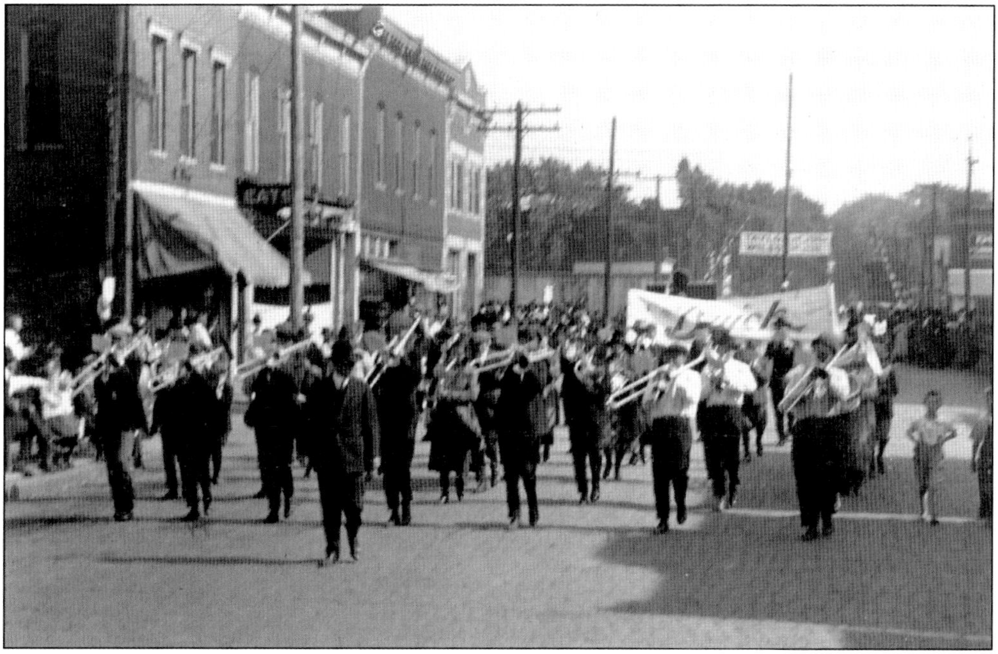

Grand marshal Thomas Langellier leads the Milford Community Band west on Jones Street toward the Dixie Highway during Milford's Dixie Days, held on October 13–14, 1920. The community celebration commemorated the official opening of the highway to traffic after paving was completed through town earlier that month. (Courtesy Iroquois County Genealogical and Historical Societies.)

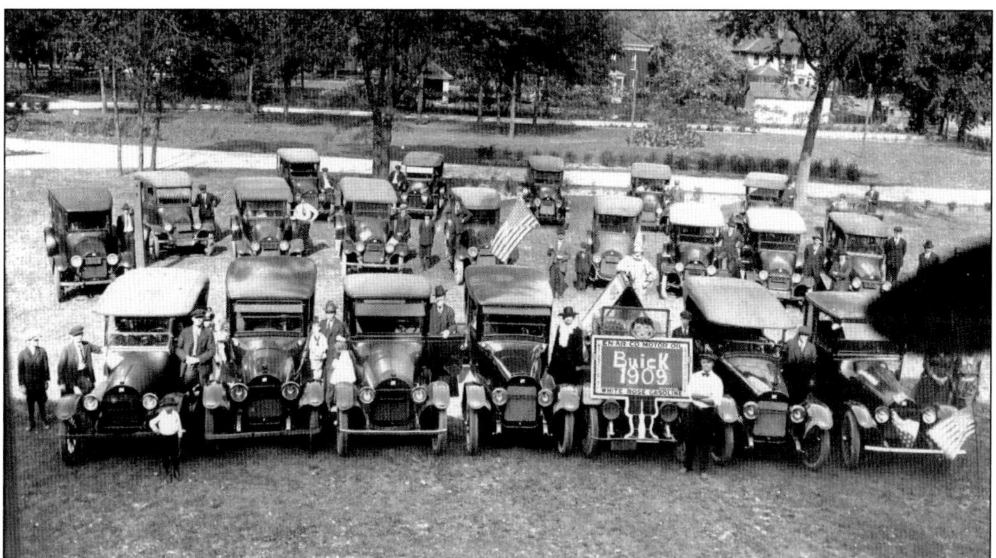

Buick dealer Louis Schaumburg poses with many of the Buick automobiles that took part in the 1920 Dixie Days parade in Milford. Schaumburg offered prizes to Buick owners that participated and lead the Buick contingent with a 1909 model that newsmen said still looked good after "many years of service." (Courtesy Fred and Frank Hines.)

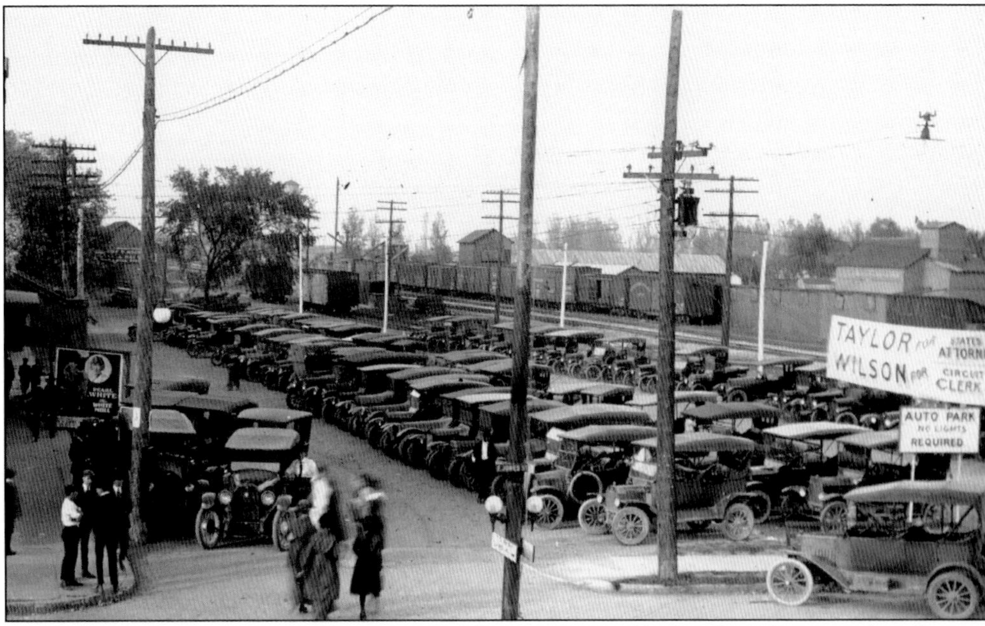

Parking was tight for the Dixie Days celebration in Milford. This photograph taken on October 14, 1920, looking north at the intersection of West and Jones Streets, shows just how popular the celebration was. No doubt many visitors that day saw silent actress Pearl White star in *The White Moll* at the village's movie theater at left. (Courtesy Iroquois County Genealogical and Historical Societies.)

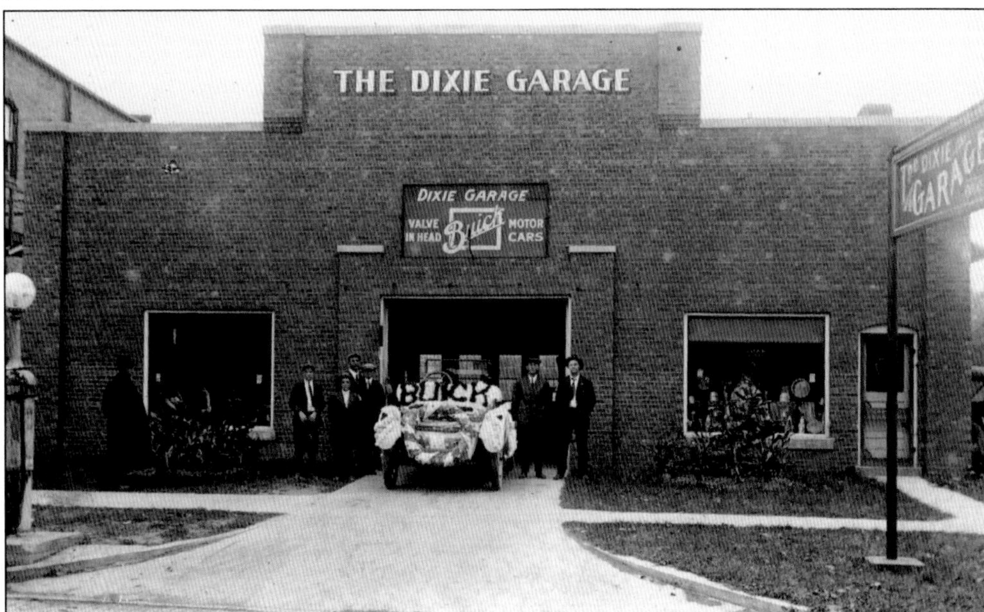

The Dixie Garage, owned by Louis Schaumburg (far right), housed his Buick dealership and service garage repairing "all makes and models." The garage was built in 1916, and the business was later taken over by his son Paul, who sold Chevrolets. Hines Automotive currently occupies the building at 1 South Chicago Street, the Dixie Highway, in Milford. (Courtesy Fred and Frank Hines)

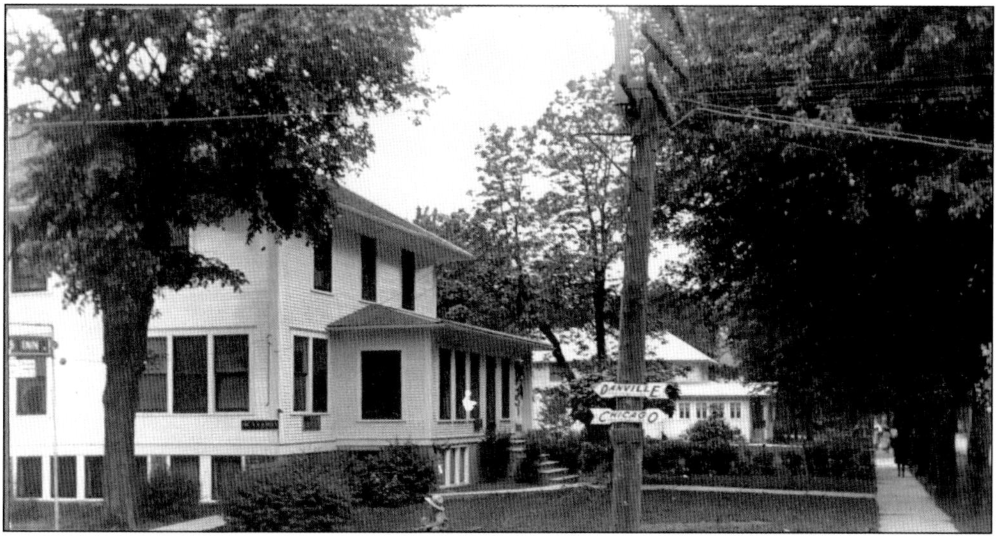

Arrows point the way to Chicago and Danville for motorists at the intersection of Chicago and Jones Streets in Milford. On the southwest corner, Dr. H. D. Junkin built the Milford Hospital in 1912. Earl Benner later ran the Traveler's Inn here with a restaurant and tavern in the basement, while his brother George, a dentist, had an office on the first floor. The building was torn down in the fall of 2008 for a new Casey's gas and convenience store. (Courtesy Iroquois County Genealogical Society.)

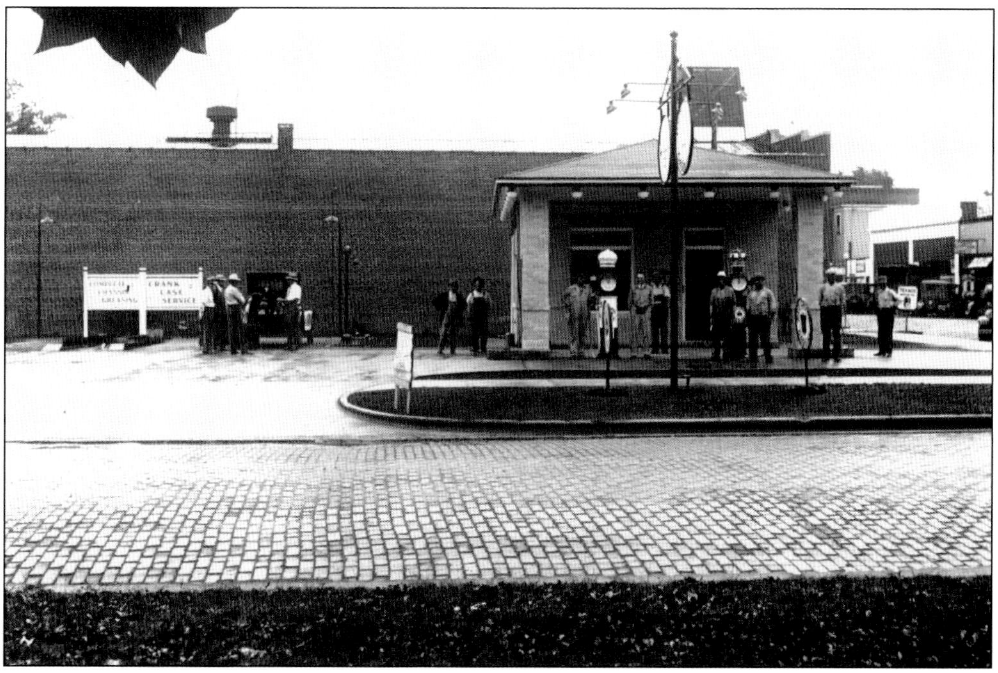

On the northeast corner of Chicago and Jones Streets in Milford, Standard Oil built this classic-style service station in the early 1930s. The station included an open-air lift (left) used for oil changes and lube jobs. The brick pavement of the Dixie Highway through Milford is visible in the foreground. Today the Dairy Queen occupies the site. (Courtesy Iroquois County Genealogical and Historical Societies.)

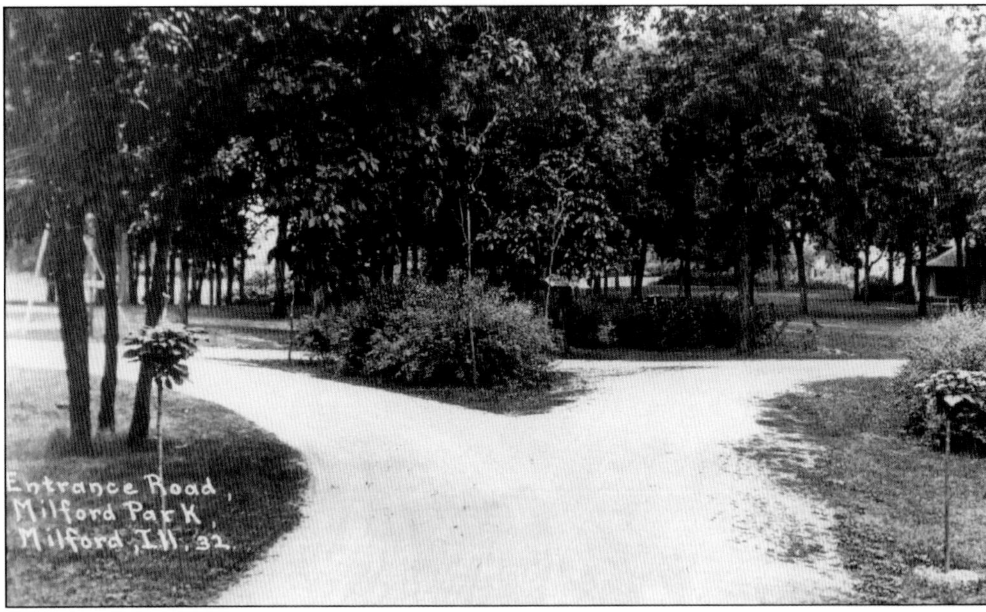

The entrance to Milford Park off Chicago Street, the Dixie Highway, was located south of the business district adjacent to the school grounds. The park offered free camping and "every convenience to the tourist," who was encouraged in advertising to "make this a point on your trip." (Courtesy Milford Public Library.)

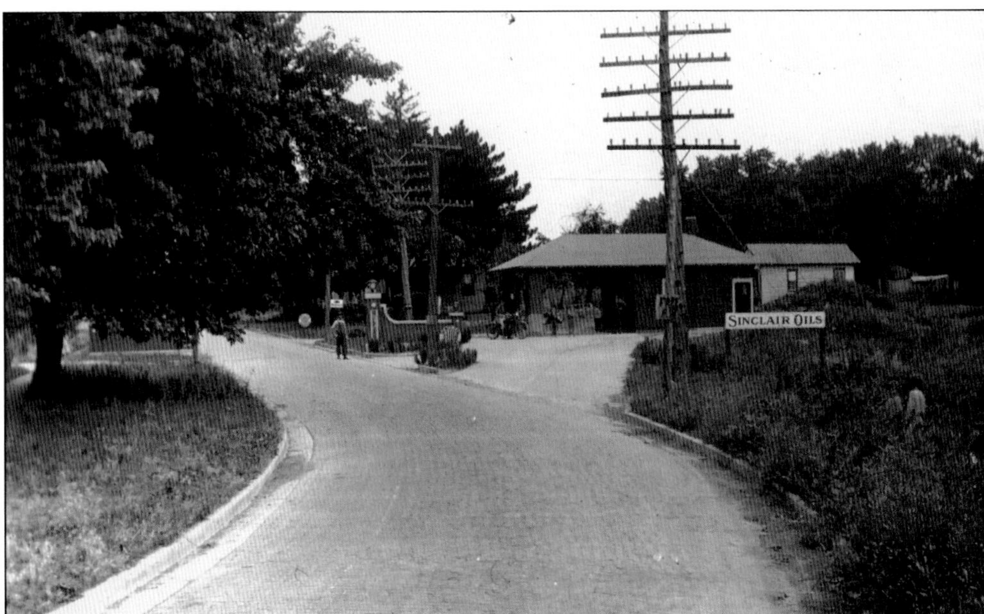

Art Thomas opened this service station under the Sinclair brand in September 1921. The *Milford News* reported he greeted northbound tourists with a "smile and howdy do." The station was located north of the south "gate" to the village. Milford citizens contributed funds for the construction of attractive brick entranceways at both ends of town on the Dixie Highway. (Courtesy Iroquois County Genealogical and Historical Societies.)

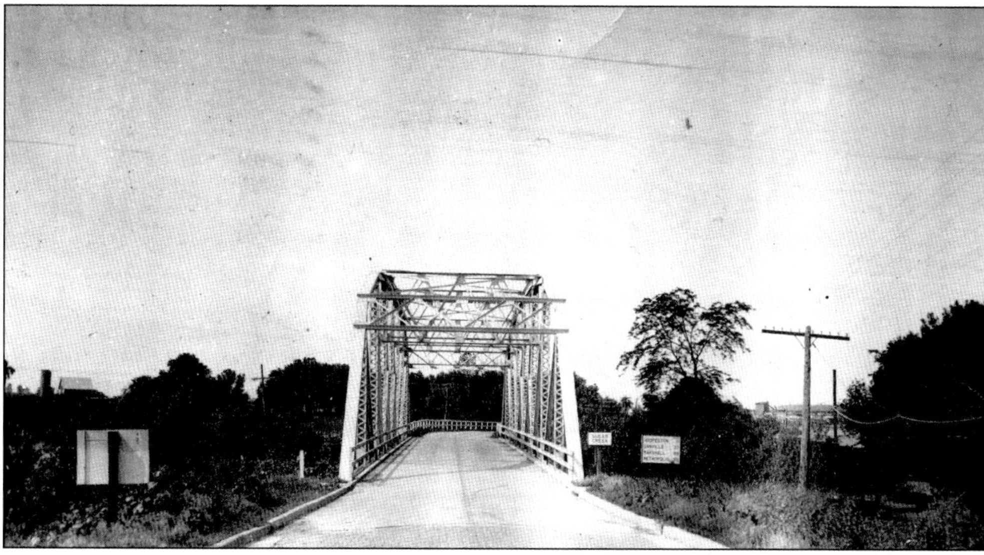

This steel girder bridge over Sugar Creek south of Milford replaced an old truss bridge with a wood plank deck built in the latter part of the 19th century. Both bridges were prone to flooding and were located west of the current alignment of Illinois Route 1, the Dixie Highway, just south of town. This view looks south, and the mileage sign at right indicates distances to Hoopeston, Danville, Marshall, and Metropolis. (Author's collection.)

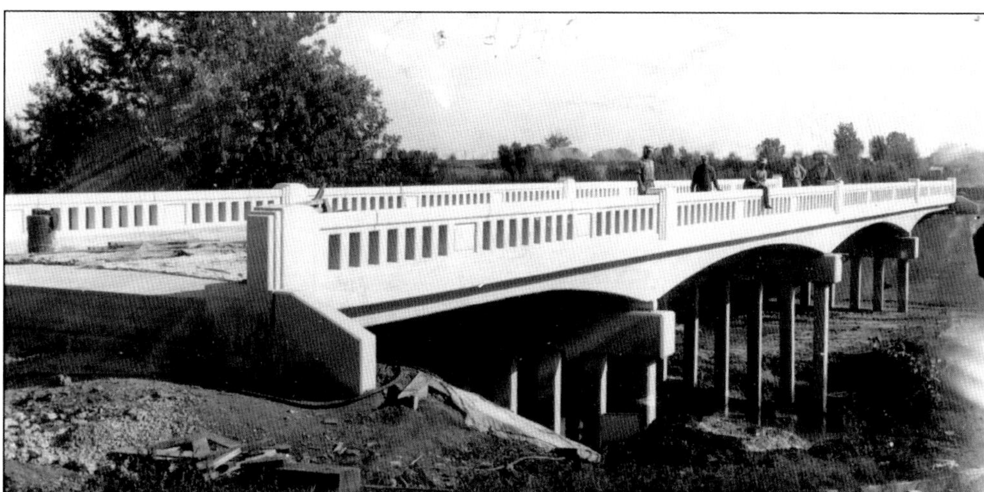

The "new" Sugar Creek Bridge was constructed in 1938 and was part of a larger program, which relocated Illinois Route 1, the Dixie Highway, to an alignment east of the original route for four and a half miles south of Milford. The new alignment eliminated two dangerous right-angle turns and saved a mile of travel. (Courtesy Iroquois County Genealogical and Historical Societies.)

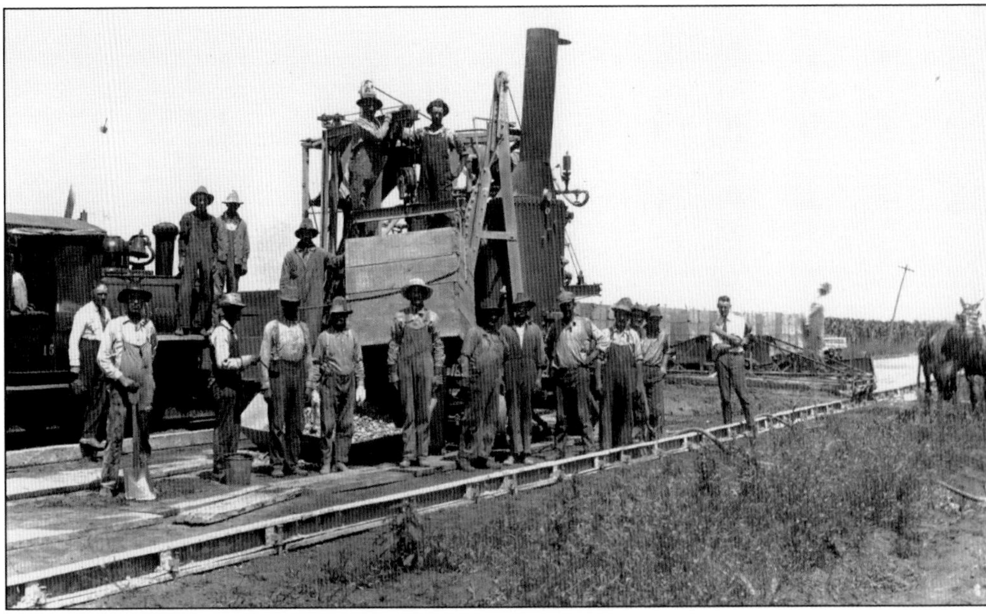

Bates and Rogers Construction work crews pause near Wellington while paving the Dixie Highway in the summer of 1920. The photograph is a great shot showing the light-rail locomotive used for hauling aggregate materials, the steam-powered paver, and a screeding or finishing device with a team of horses next to it. (Courtesy Iroquois County Genealogical and Historical Societies.)

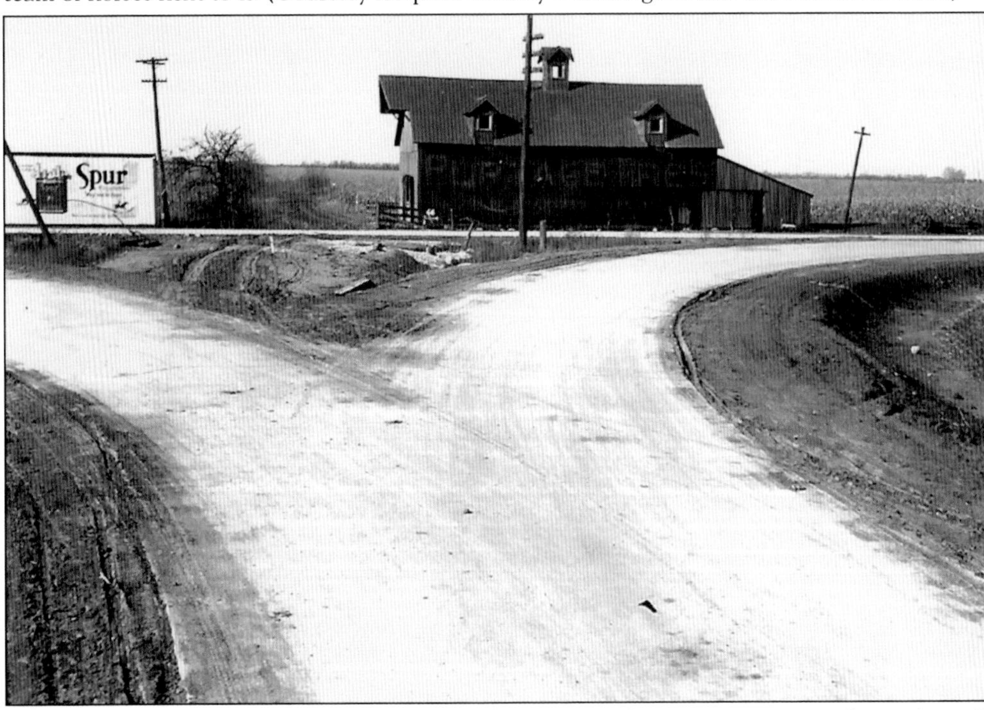

The intersection of the Wellington Road and the Dixie Highway was constructed as a classic Y popular in early road-building efforts. Vintage advertising for Spur cigarettes and Portage tires add interest to this scene just west of Wellington. Today the intersection is a more conventional T configuration. (Courtesy Iroquois County Genealogical and Historical Societies.)

Five

Vermilion County

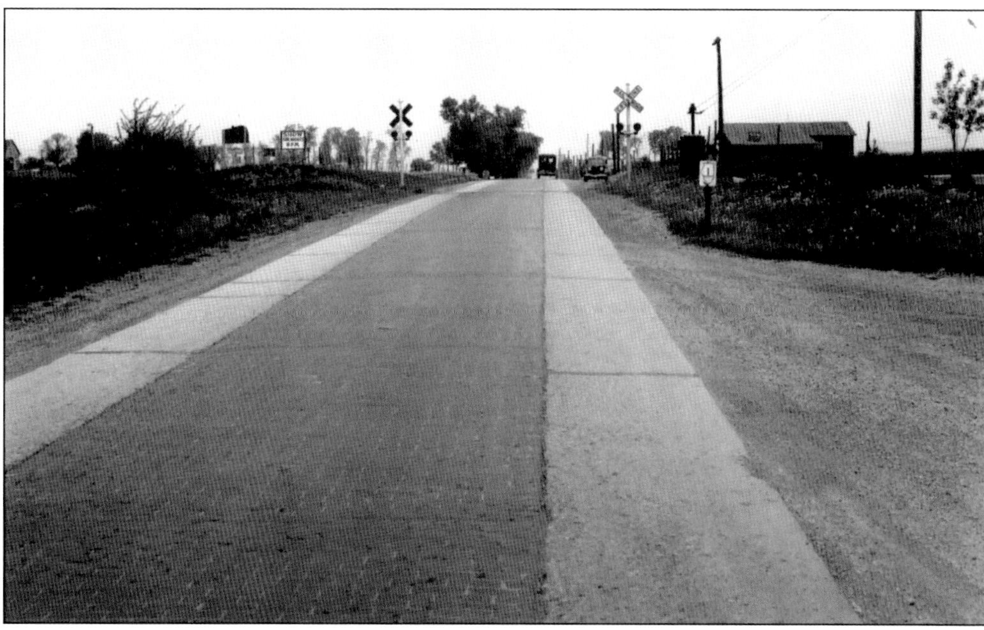

Paving the Dixie Highway from the Iroquois-Vermilion County line through and south of Hoopeston was completed in 1917. Vermilion County officials chose to pave the highway with brick to a width of 10 feet. This view looking north on the Dixie Highway in Hoopeston at the New York, Chicago and St. Louis Railroad crossing from 1940 shows the four-foot concrete shoulders added to the highway in the 1920s to make travel safer. (Courtesy Illinois Department of Transportation.)

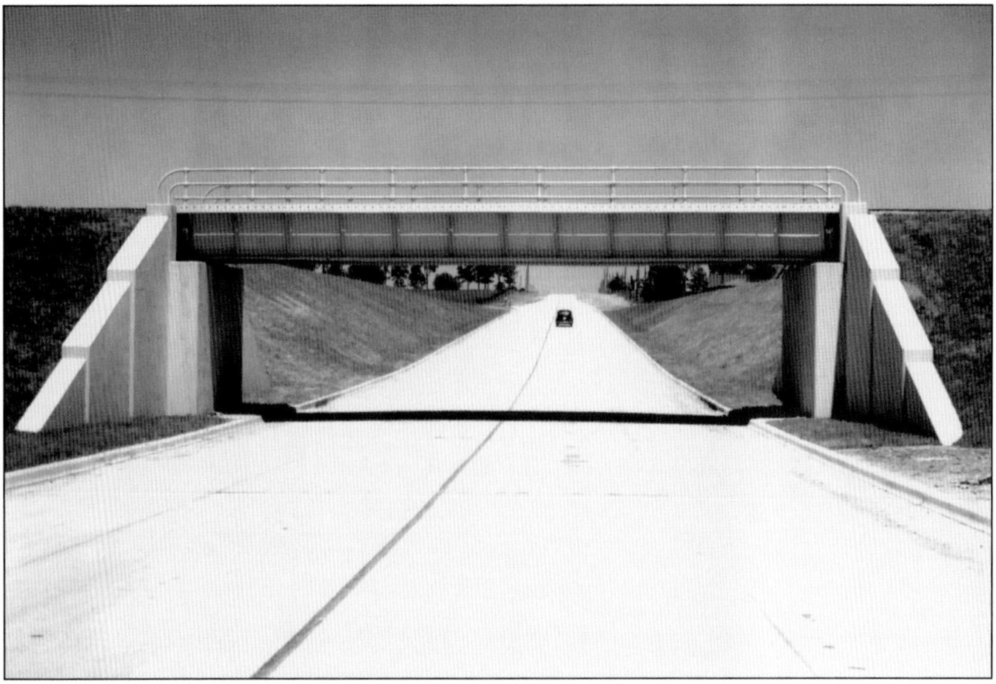

As rail crossings were a major hazard for motorists, the State of Illinois completed an underpass in Hoopeston at the New York, Chicago and St. Louis Railroad crossing just north of Main Street on Illinois Route 1, the Dixie Highway, in 1941. After the railroad tracks were abandoned, the trestle was removed in 1990. (Courtesy Illinois Department of Transportation.)

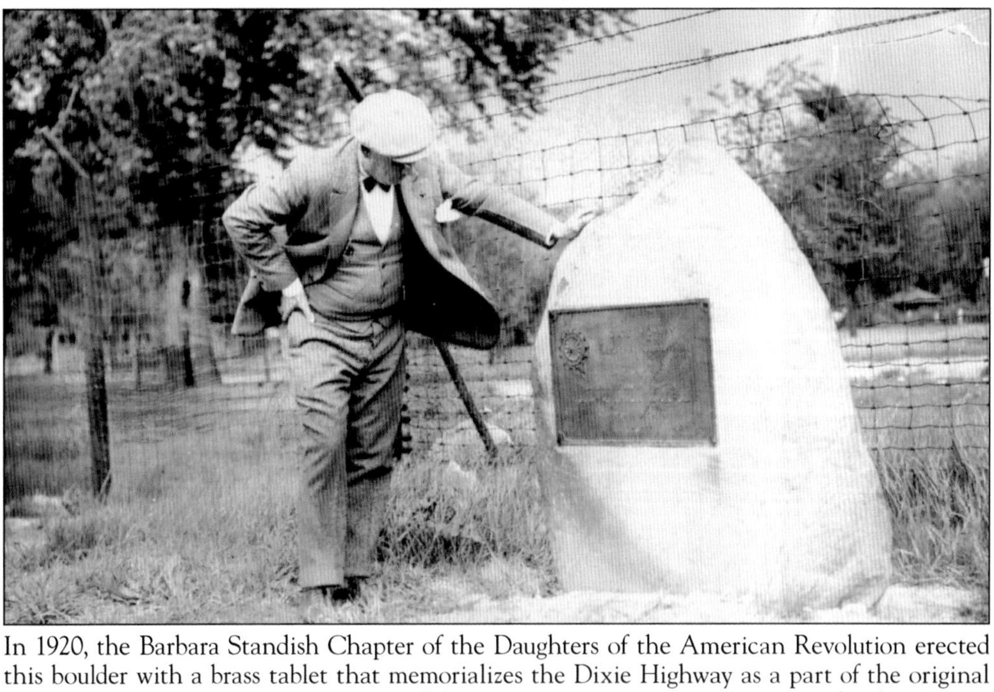

In 1920, the Barbara Standish Chapter of the Daughters of the American Revolution erected this boulder with a brass tablet that memorializes the Dixie Highway as a part of the original Hubbard's Trail. The monument can still be seen in Hoopeston on the west side of McFerren Park, just east of the highway. (Courtesy Chicago History Museum.)

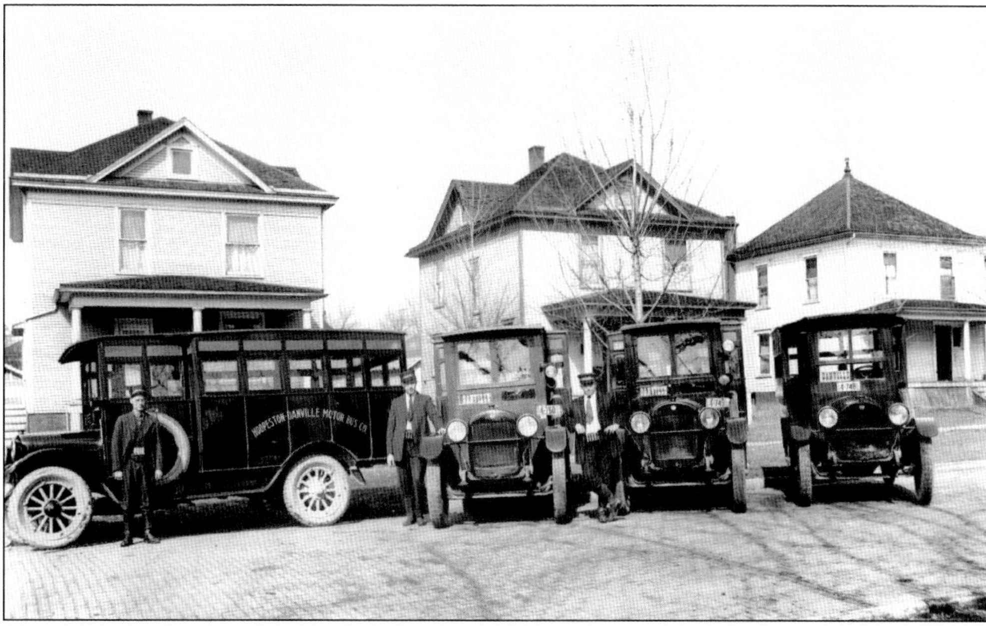

Lester Moore (left) stands with other drivers of the Hoopeston-Danville Motor Bus Company in this photograph from 1922 taken on Main Street in Hoopeston. The bus company ran a route down Illinois Route 1, the Dixie Highway, through Rossville to downtown Danville, providing a convenient way for workers, shoppers, and others to get to and from the city. (Courtesy Hoopeston Public Library.)

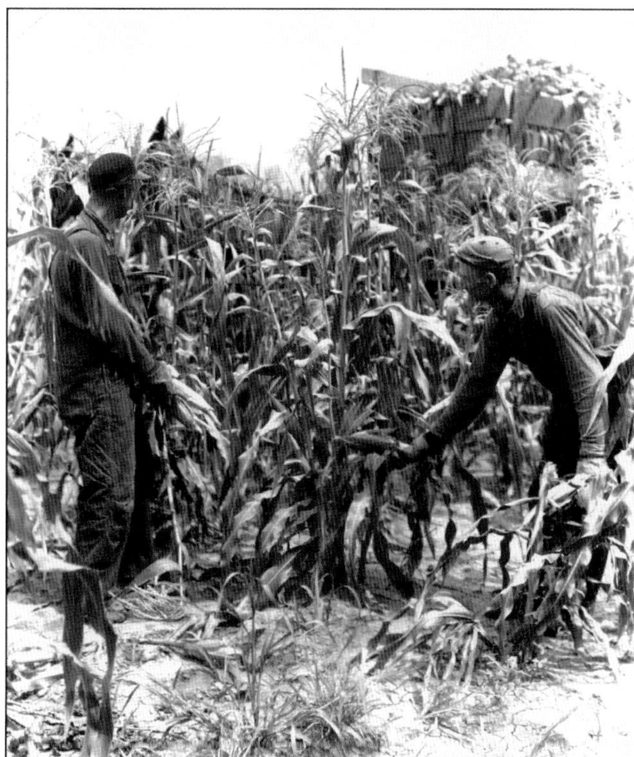

Hoopeston was long known as "the Sweet Corn Capital of the World." For years, sweet corn ears were picked by hand. Laborers who pulled the ears from the stalks were affectionately known as "cornjerkers." Although machines now harvest sweet corn and the area is no longer a major sweet corn producer, Hoopeston Area High School athletic teams are still named the Cornjerkers, and the town still sponsors an annual sweet corn festival. (Courtesy Hoopeston Public Library.)

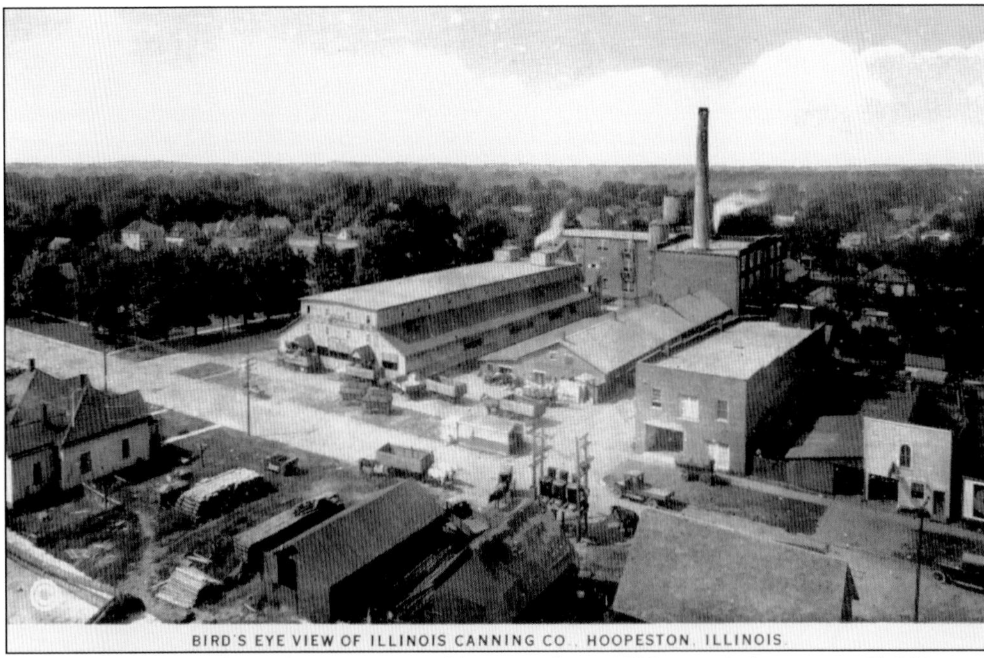

A Hoopeston Chamber of Commerce publication from 1930 stated there were 22,900 acres of farmland in the immediate vicinity of Hoopeston dedicated to sweet corn production. Canneries, like the Illinois Canning Company, established in 1878 and producer of Joan of Arc brand products, were once important industries and major employers in town. (Courtesy Carol Hicks.)

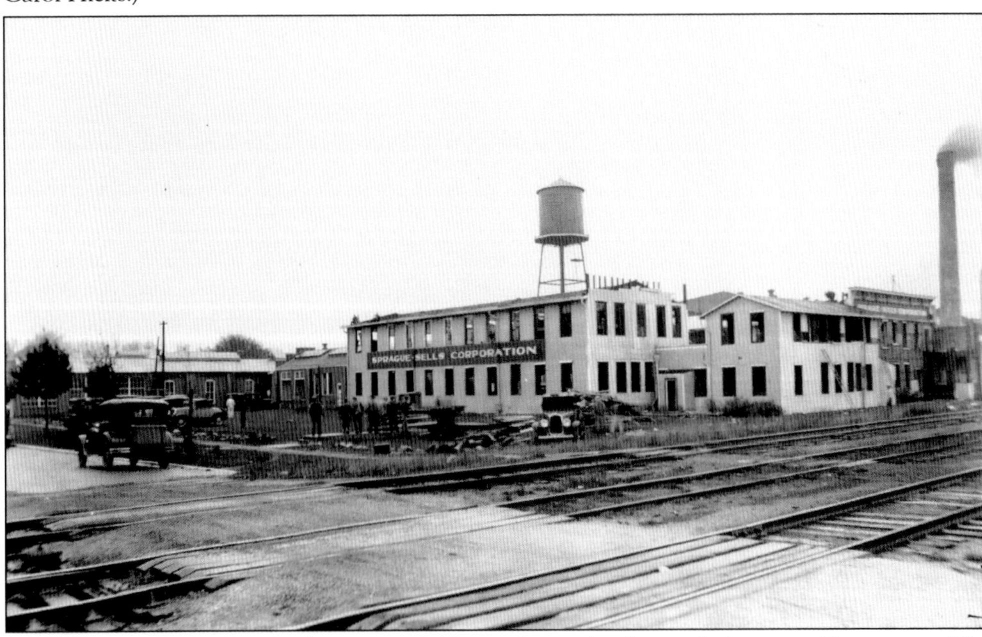

In addition to canneries, industries supporting them also located plants in Hoopeston. The American Can Company built a factory in town to produce cans, as did the Sprague-Sells Corporation, which manufactured canning machinery. Sprague-Sells later became a part of the Food Machinery Corporation (FMC). (Courtesy Hoopeston Public Library.)

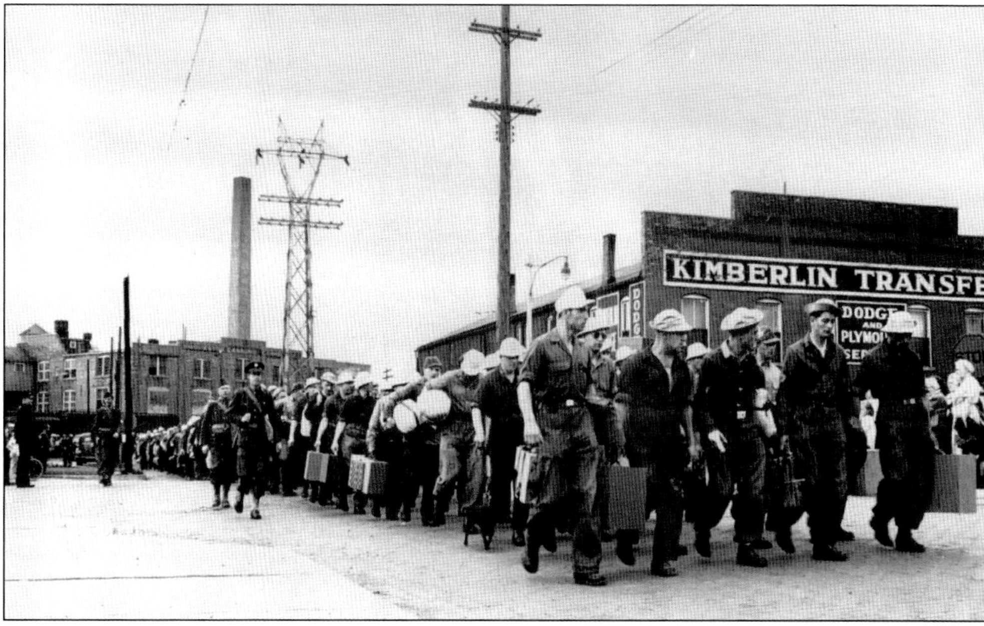

During World War II, severe labor shortages affected the factories in Hoopeston. To alleviate the shortage, German prisoners of war were sent to the community to help with the harvest and in local canneries. A camp was set up east of the Dixie Highway, at Second Avenue and Illinois Route 9. Here prisoners march to the camp after disembarking from a train in front of the Hoopeston Canning Company. (Courtesy Hoopeston Public Library.)

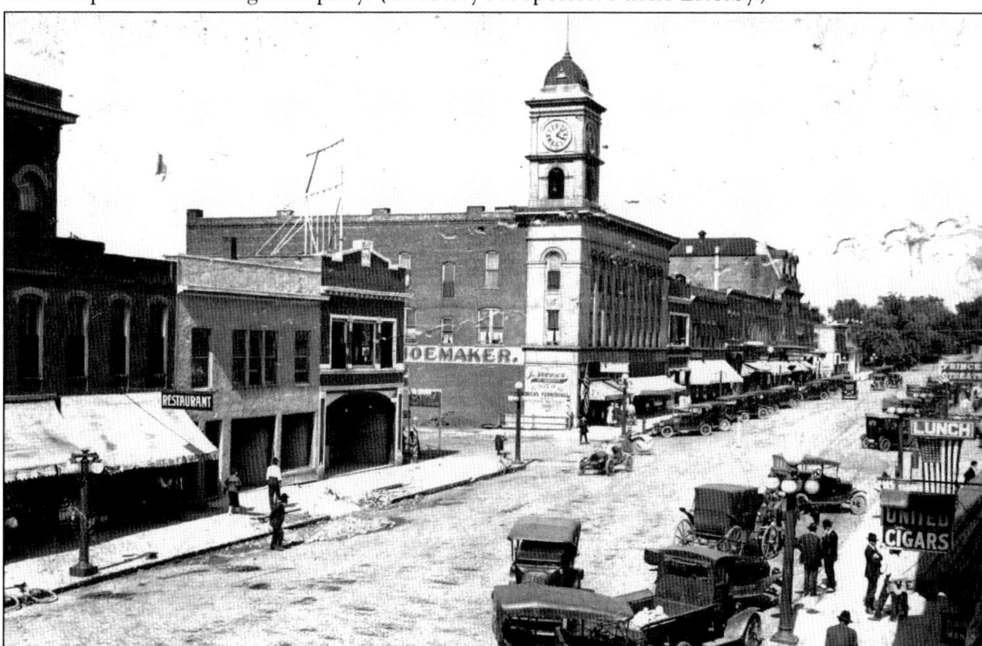

Main Street in Hoopeston was a thriving business district that offered every amenity to residents and visitors alike. The central business area of the community was located several blocks east of the Dixie Highway, centered around the C&EI tracks. This view looking east on Main Street from the rail tracks was taken in the early 1920s. (Courtesy Hoopeston Public Library.)

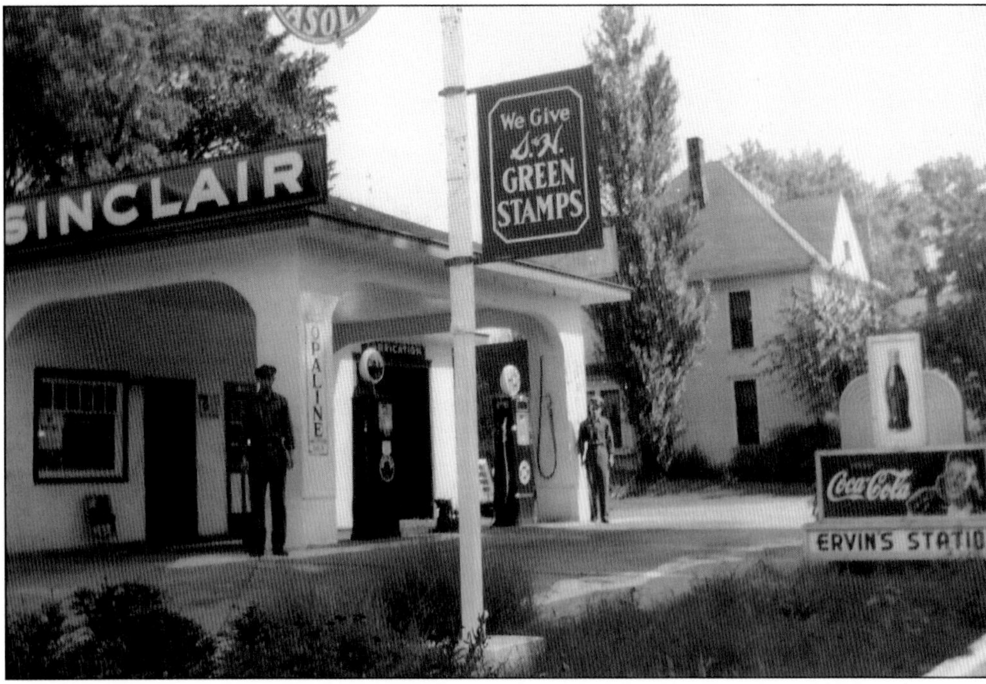

Ervin's Sinclair station in Rossville prospered with increased traffic along North Chicago Street, the Dixie Highway. The station was located on the corner of Benton Street and had an interesting art deco–style Coca-Cola advertising sign. A gasoline station was in continuous operation on this corner from the 1930s until the 1990s (Courtesy Rossville Historical Society.)

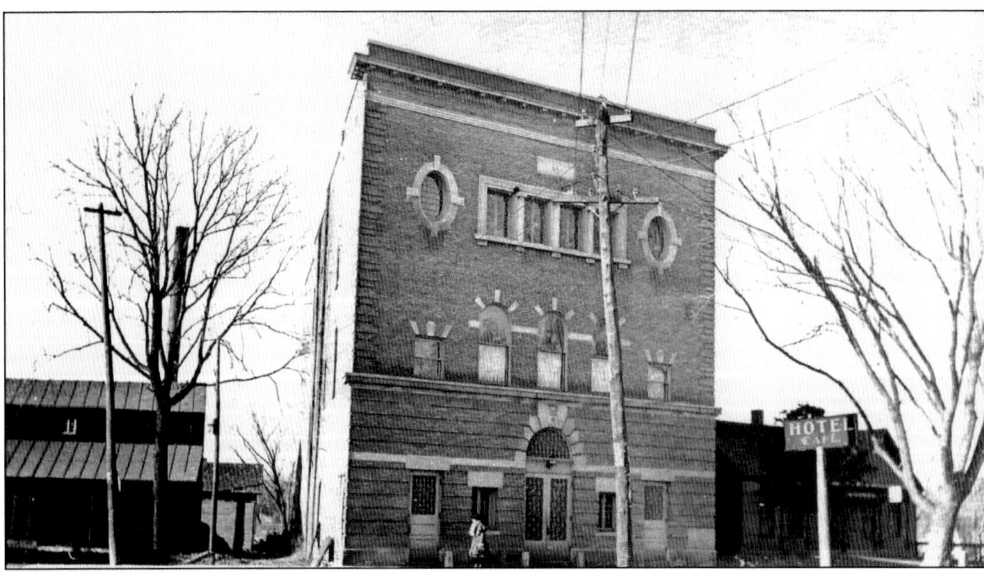

One of the most imposing structures on the Dixie Highway in Rossville, the opera house and Masonic hall building, was built in 1903 and is located at 108 North Chicago Street, just north of the Attica Road intersection. It was used for meetings and lodge functions by the Masons and as a venue for local entertainment. The opera house closed in 1931. Today the building houses an antique store and apartments. (Courtesy Wood Collection, Rossville Historical Society.)

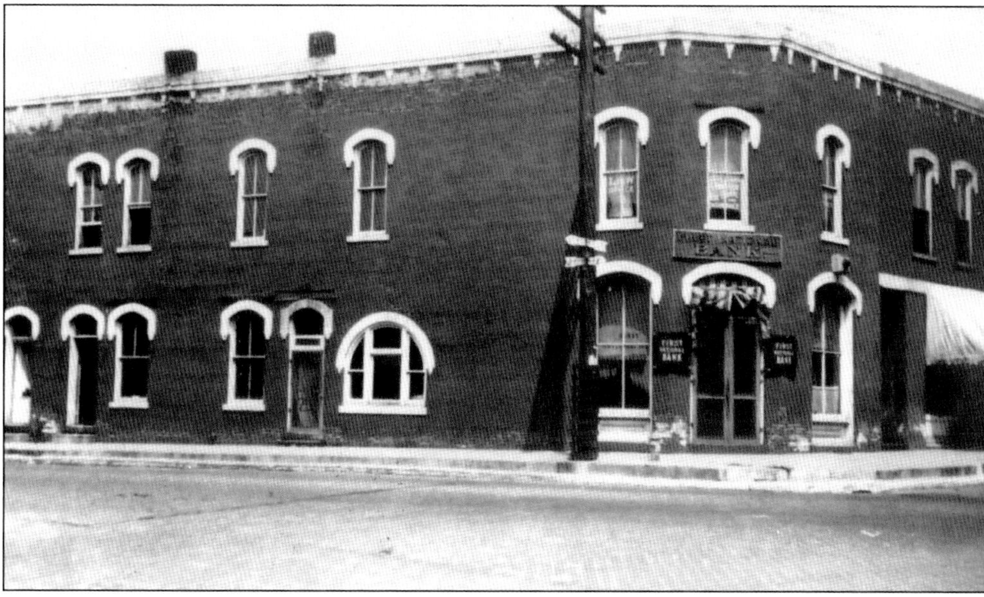

The First National Bank in Rossville was located at the northeast corner of Chicago Street, the Dixie Highway, and Attica Road, the main intersection in town. Paving of the Dixie Highway was complete when this photograph was taken. Arrow signs on the telephone pole in front of the bank point the way to Chicago, left, and Danville, right. (Courtesy Rossville Historical Society.)

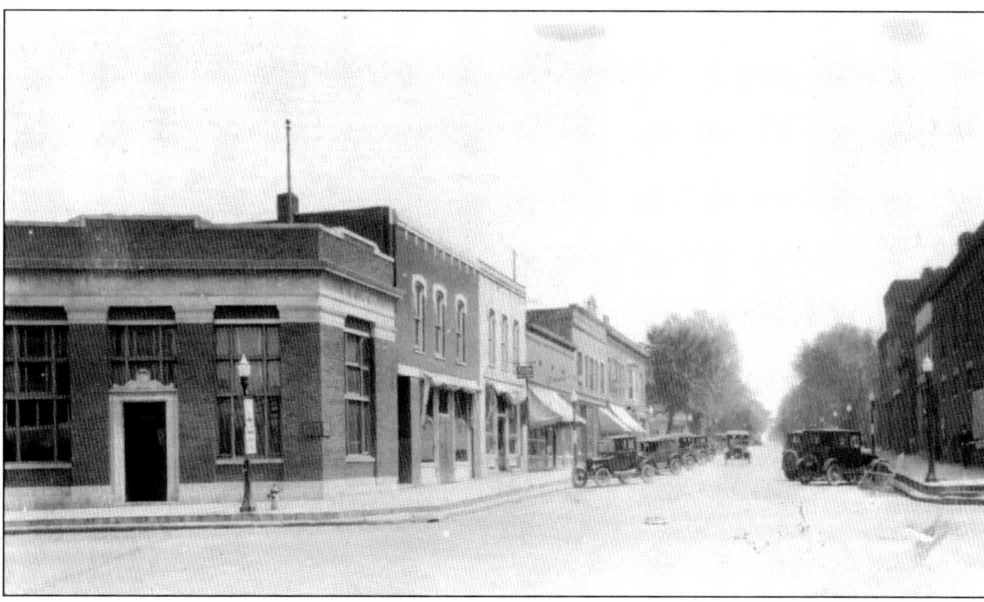

The original First National Bank building in Rossville was replaced with a new structure built on the same site in 1923. The bank, left, served as the anchor of the business district in the village at the intersection of Chicago Street and Attica Road. This view is looking east on Attica Road from Chicago Street, the Dixie Highway. (Courtesy Rossville Historical Society.)

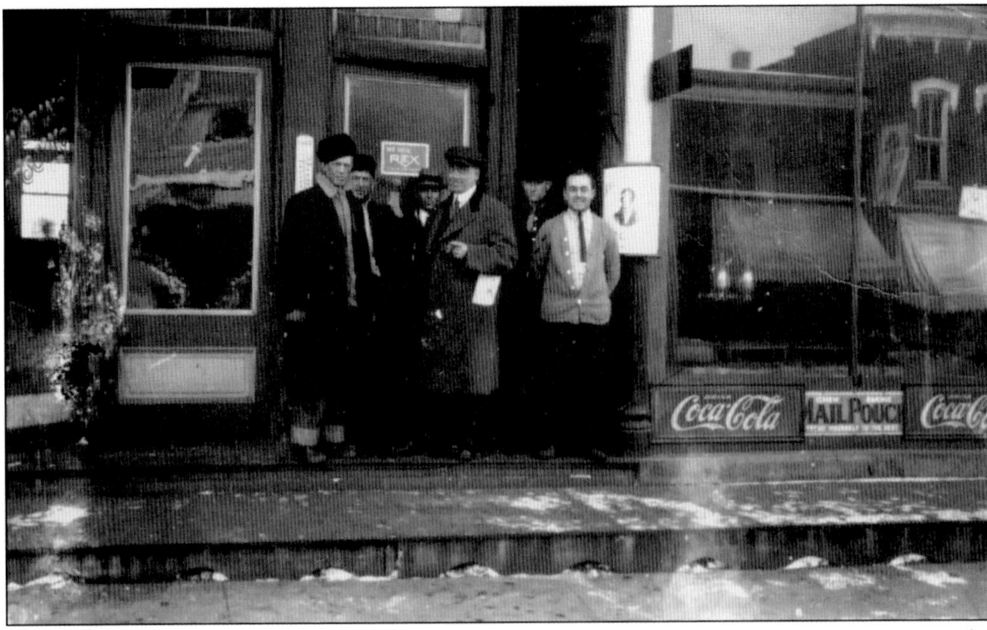

The Prather Brothers Confectionary store, located at 104 South Chicago Street, was a popular gathering spot in Rossville. Pictured in the photograph are Albert Prather, Tom Neves, Everett Lee, Harvey Bracken, and "Brownie" Prather. The building, located on the west side of the Dixie Highway, along with a number of others, was destroyed by fire in February 2004. (Courtesy Rossville Historical Society.)

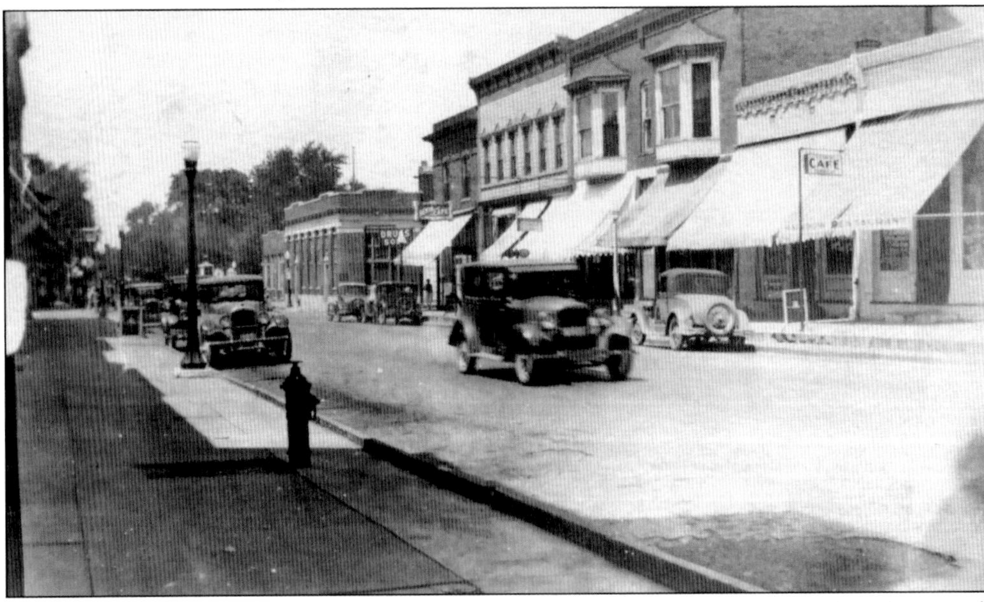

Traffic was light on Chicago Street, the Dixie Highway in Rossville, when this photograph was taken in 1930. The view is looking north toward the Attica Road intersection. The Dixie Café, located just south of the intersection, was one of the first businesses to capitalize on its location when owner C. W. Williamson renamed his restaurant after the highway was routed through town. The Dixie Café operated under this name until 1945. (Courtesy Rossville Historical Society.)

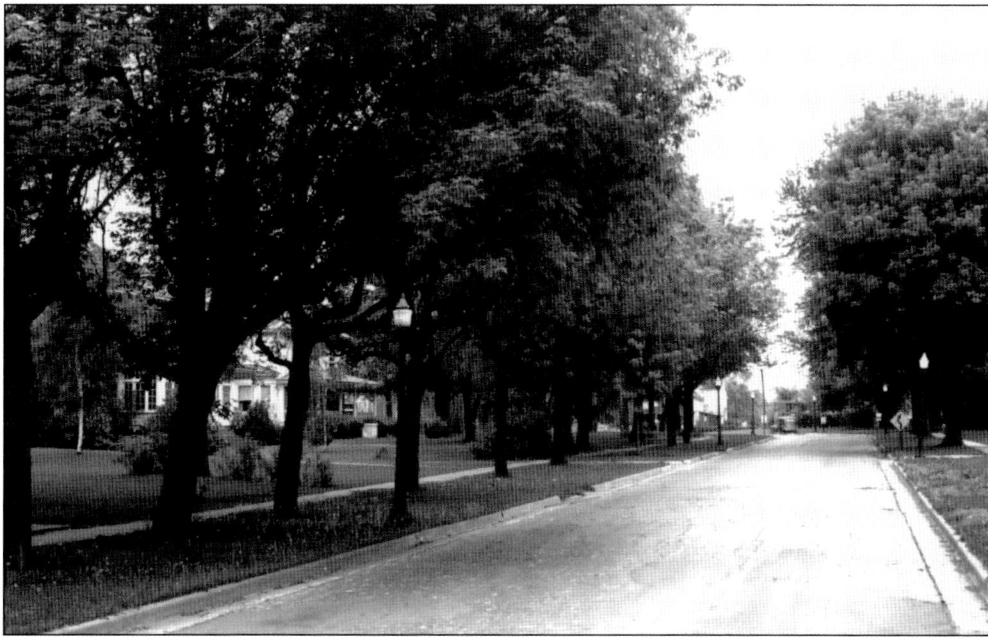

As early as 1932, state highway engineers worked on plans to eliminate a dangerous S curve in Rossville on the Dixie Highway near Gilbert Street. The curve was the scene of many accidents even though it was located on a residential stretch of the highway in town. It would take another eight years before the curve would be straightened. This view from May 1940 is looking north on Chicago Street, the Dixie Highway, toward Gilbert Street. (Courtesy Illinois Department of Transportation.)

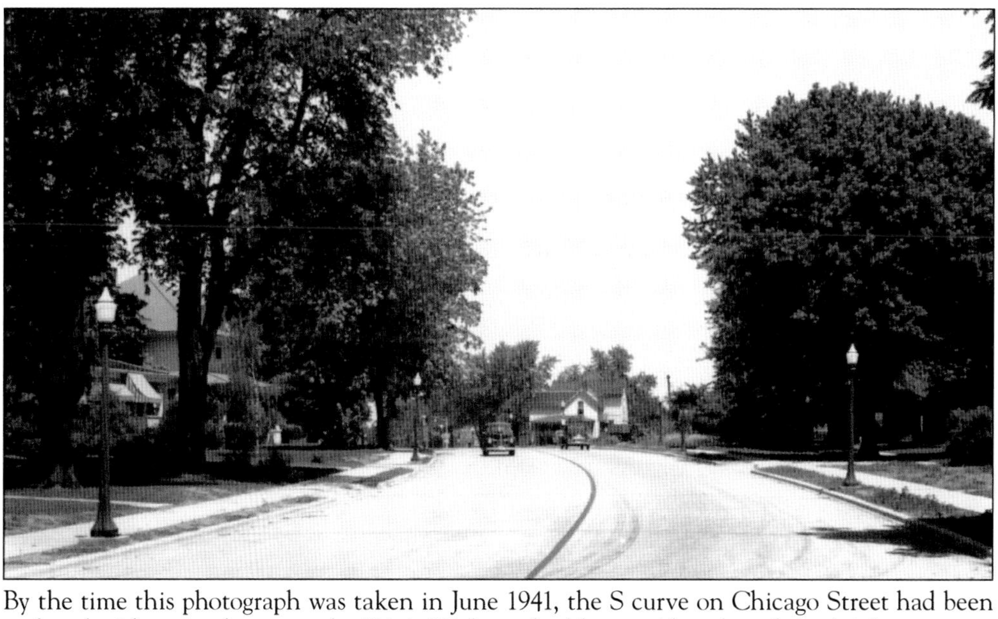

By the time this photograph was taken in June 1941, the S curve on Chicago Street had been replaced with a simple curve, the Dixie Highway had been widened, and the brick pavement was replaced with concrete. A close review of this and the previous photograph shows how the parkway was narrowed and trees removed to accommodate these improvements in Rossville. (Courtesy Illinois Department of Transportation.)

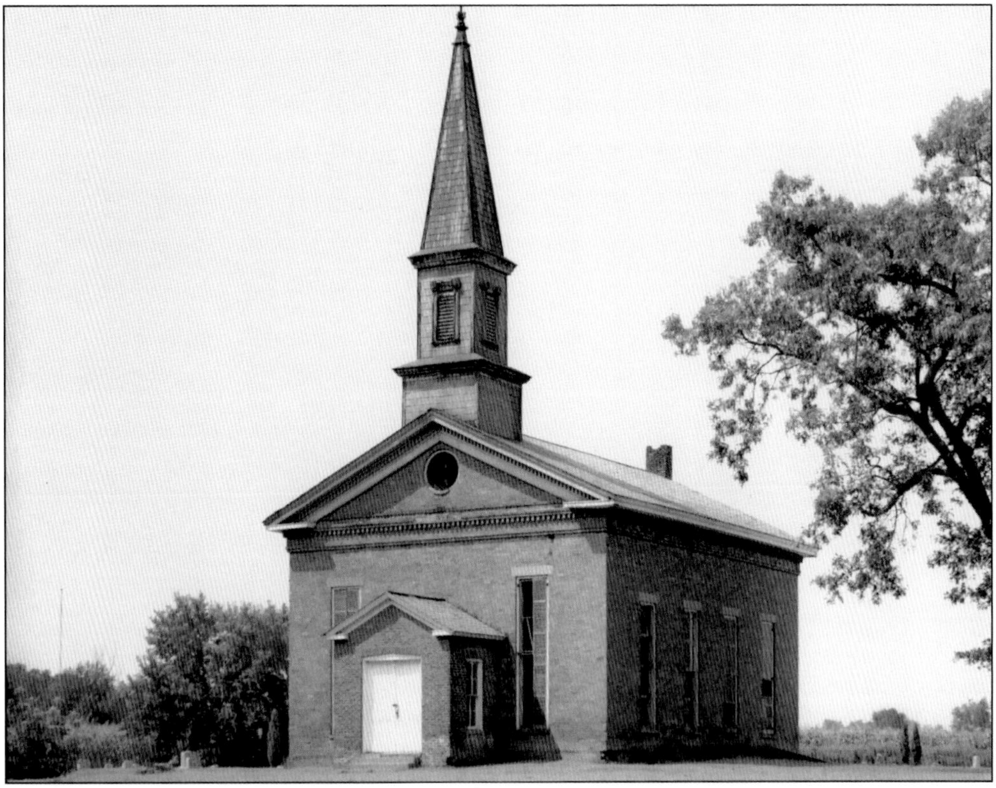

An original Hubbard's Trail stone mile marker still stands on the east side of Illinois Route 1, the Dixie Highway, several miles south of Rossville along the North Fork of the Vermilion River. The marker is located west of Mann's Chapel, a local landmark, built in 1857, and the oldest surviving church structure in Vermilion County. (Courtesy Kevin Young.)

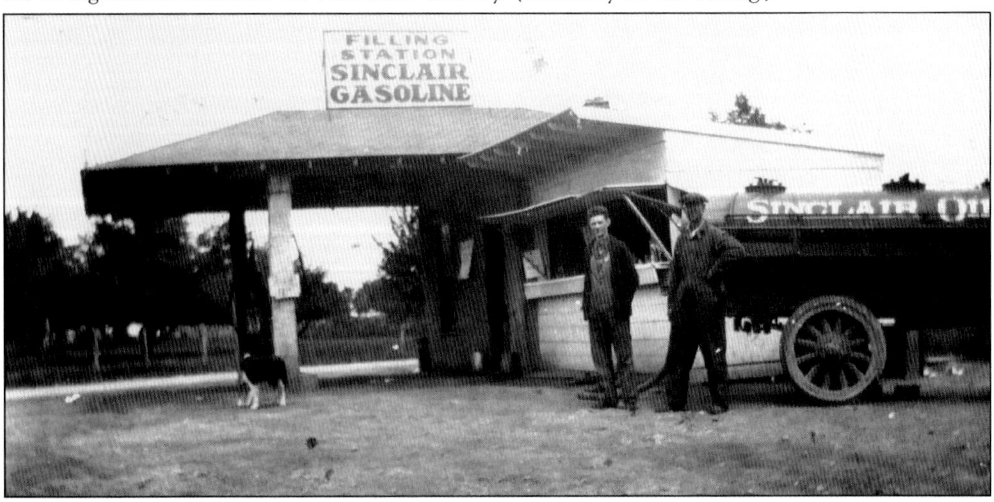

Paul Clem (left) watches as a tanker truck fills the underground gas tanks at the Sinclair station at Moore's Corner, the southeast corner of the Dixie Highway and the Bismarck-Jamesburg Road. Typical of most of the filling stations along the route, the station had a couple of pumps and a small food stand attached offering a fast meal and cool drink to travelers. (Courtesy Norma Clem Bailey.)

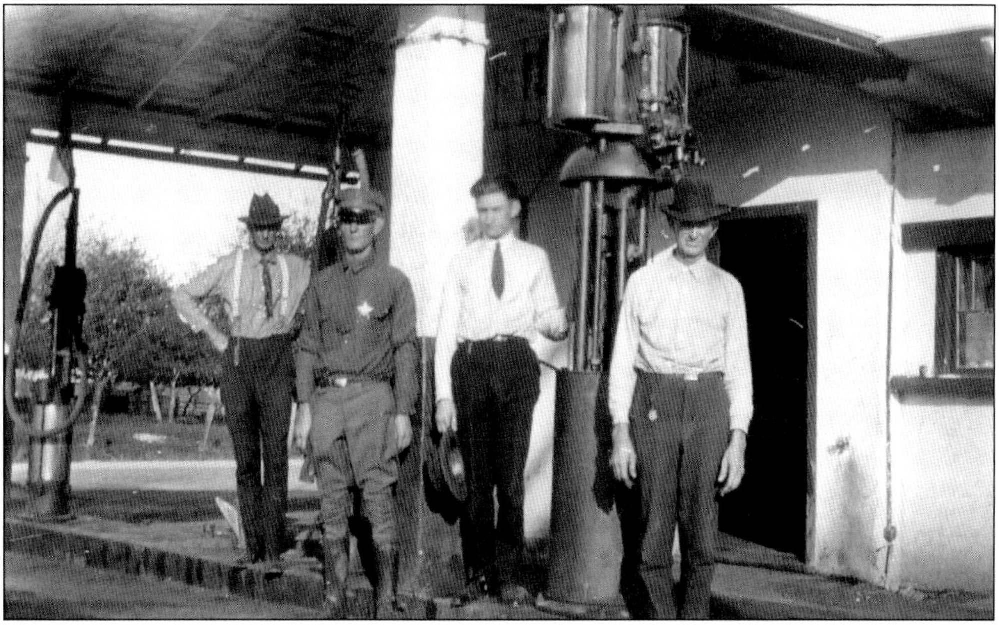

Illinois state trooper Robert Fisher poses for this photograph with Paul and Albert Clem on his left at the Moore's Corner Sinclair station. Fisher was well known, patrolling the Dixie Highway and other roads in Vermilion County. Fisher was killed while on duty on April 24, 1929, near Rossville when a bus lost control on the highway and struck his motorcycle. (Courtesy Norma Clem Bailey.)

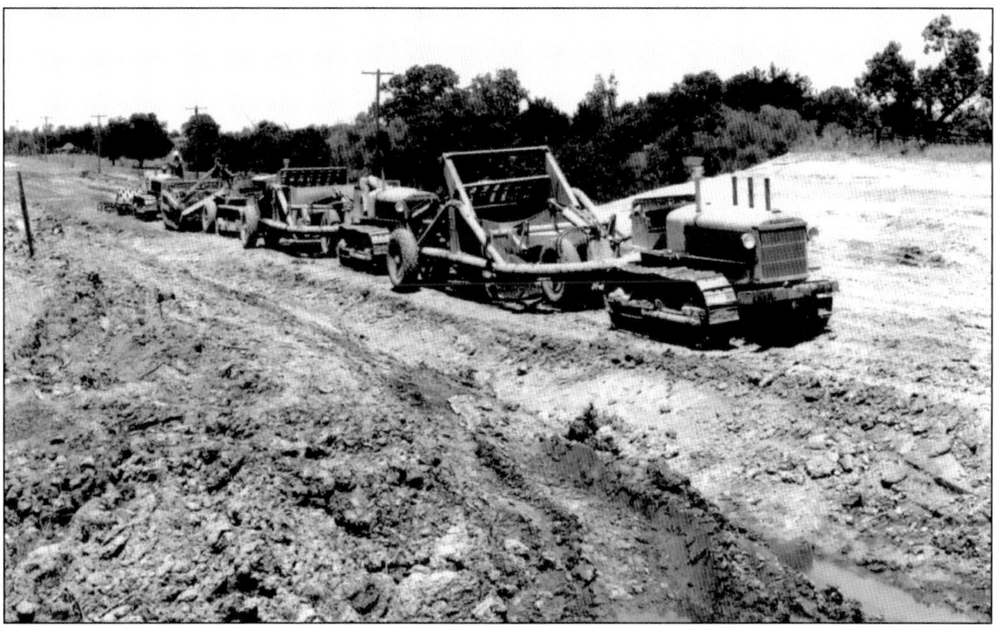

North of Danville, just north of the North Fork of the Vermilion River, the Dixie Highway and Illinois Route 1 traveled along the brick pavement of what is now Old Seaton Hill Road. Road-grading crews worked during the summer of 1938 on this new alignment of the highway just east of the original pavement to make the hill less steep for traffic. (Courtesy Illinois Department of Transportation.)

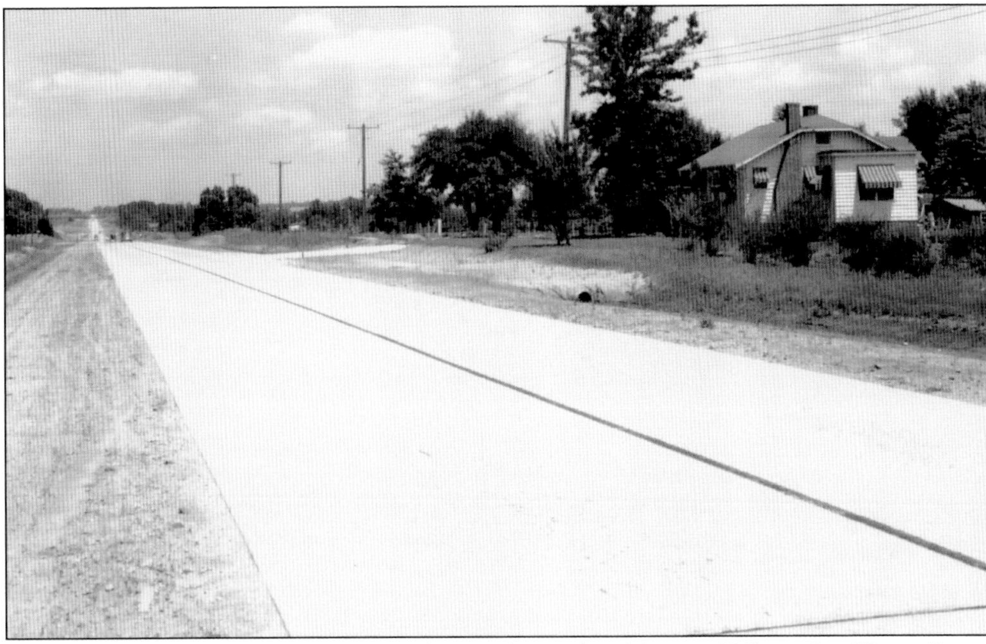

New concrete pavement and the cut reducing the grade of Seaton Hill north of Danville are visible in this photograph from June 1939. The original route of the highway, which retains much of its brick pavers, is located just west of the current pavement and is now a private drive serving residences along the road. (Courtesy Illinois Department of Transportation.)

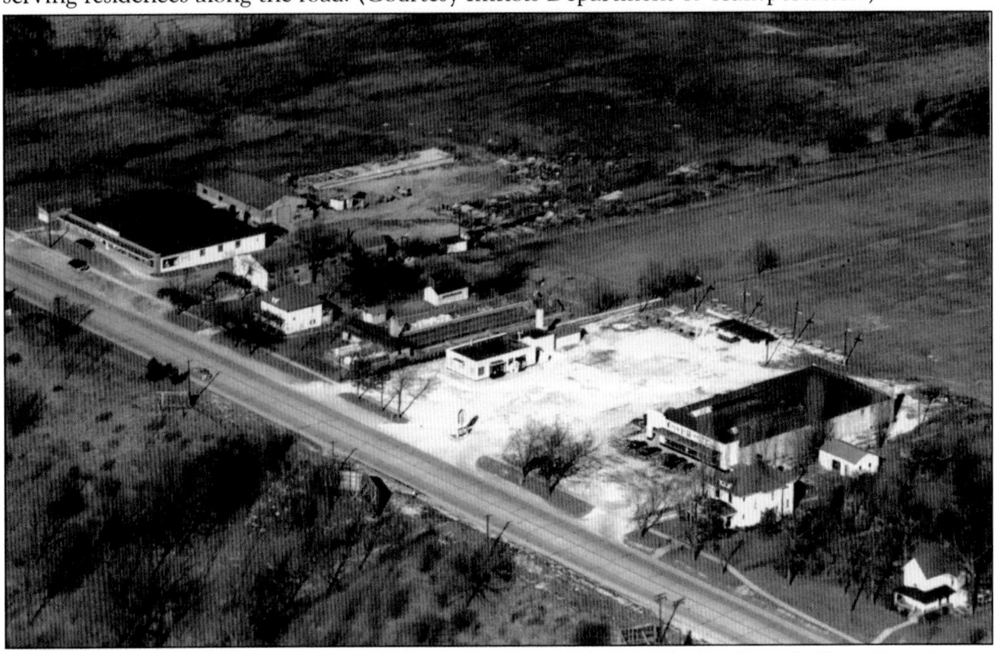

Considered "out in the country," the area in the 2700 and 2800 block of North Vermilion Street, the Dixie Highway in Danville, began to experience commercial development by the late 1940s. Pictured are, from left to right, Goodloe E. Moore, Inc., Wright's Florist, the Crest Restaurant, and the Grab-It-Here Grocery Store. Today both sides of this now four-lane stretch of roadway are filled with retail and dining establishments. (Courtesy Vermilion County Museum.)

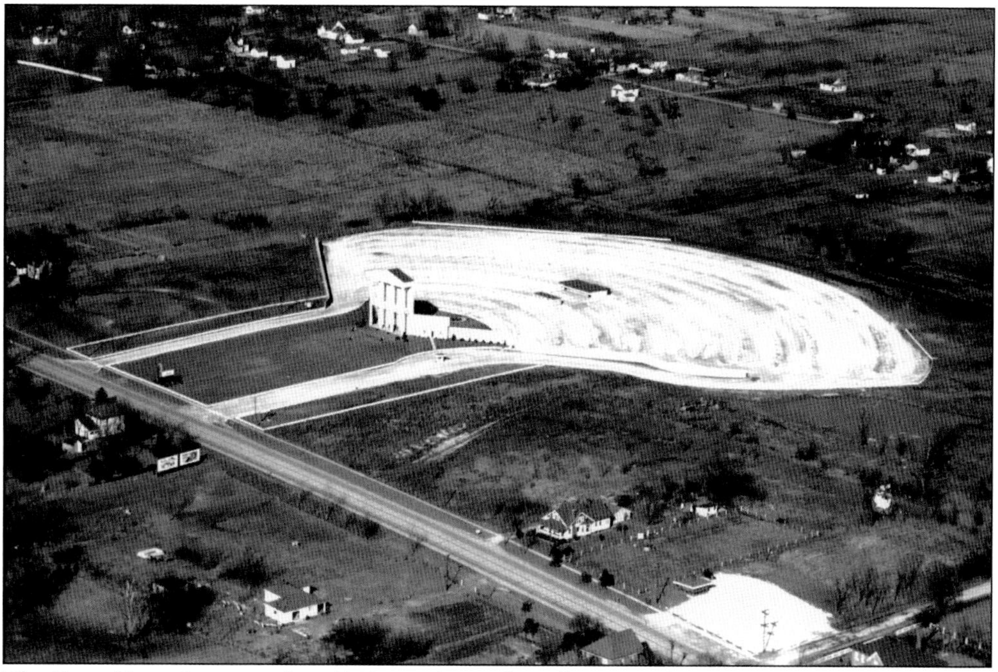

The Dixie Drive-In, opened in 1949, was located on the east side of the Dixie Highway/Illinois Route 1, north of Liberty Lane in Danville. The street side of the screen was built to mimic the facade of a southern mansion. Initially built to accommodate 700 cars, the drive-in later expanded to three screens. It succumbed to reduced attendance and closed in 1983. It was torn down in 1985 and has been replaced by a car wash, apartment building, and other commercial development. (Courtesy Vermilion County Museum.)

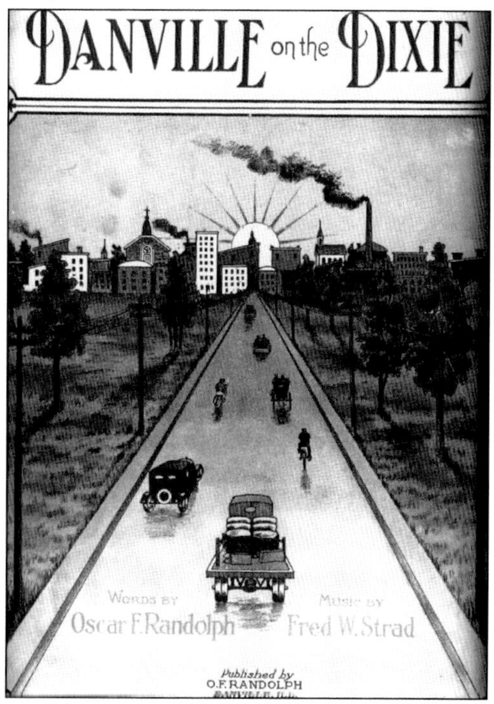

City fathers worked hard to get the Dixie Highway to come through Danville. It was the southernmost point on the highway in Illinois, and Danville residents were great boosters of the road. Local residents Oscar Randolph and Fred Strad composed the tune "Danville on the Dixie," which was published with this fanciful cover showing the highway entering the city. (Courtesy Vermilion County Museum.)

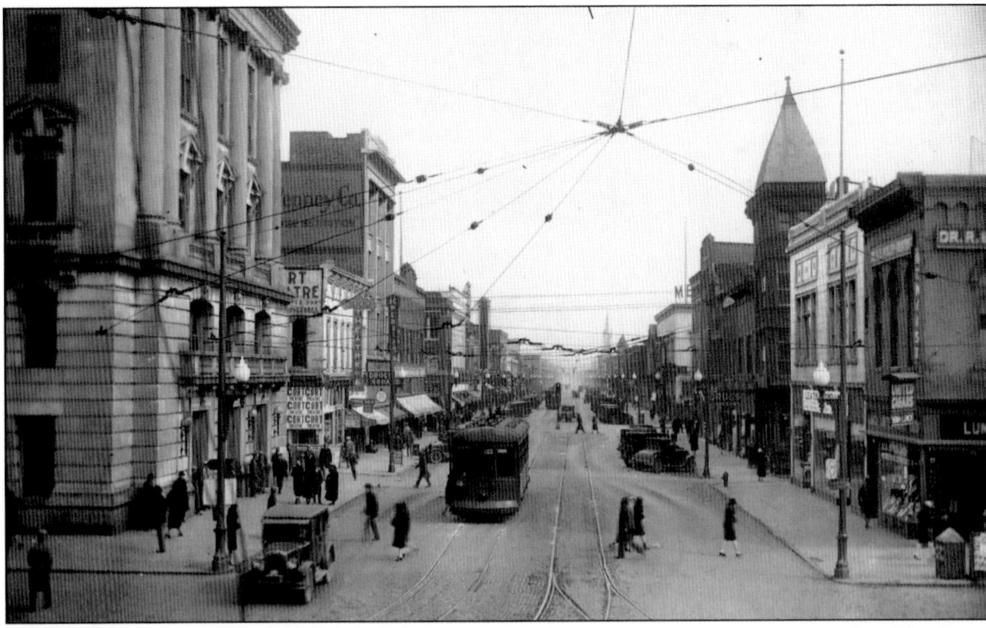

The intersection of Vermilion and Main Streets was the hub of the city of Danville. The Dixie Highway traveled along Vermilion Street to Main Street, then east to the Indiana state line. This view shows Main Street looking east from Vermilion Street. An arrow sign on the streetlight pole on the right points travelers to Indianapolis. Another sign also identifies Main Street by its early designation as Illinois Route 10. (Courtesy Vermilion County Museum.)

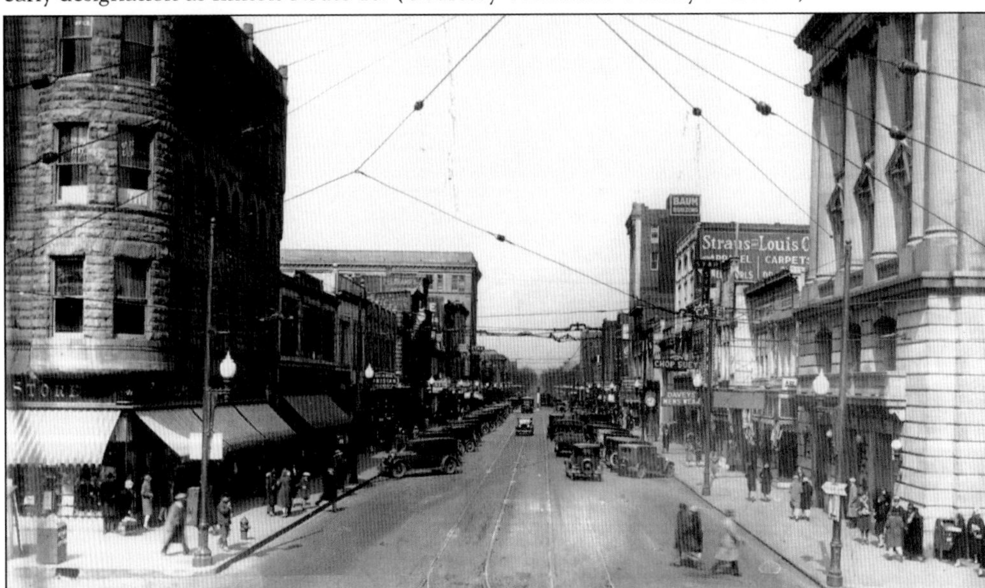

Vermilion Street north of Main Street in Danville was full of activity when this photograph was taken just after 9:00 a.m. according to the clock on the sidewalk in the center of this picture. Angle street parking narrowed the flow of traffic and was later abandoned in favor of parallel parking. The interurban tracks were removed by the early 1950s, which also allowed for traffic to flow more freely. An arrow sign on the street pole at right proclaims, "Chicago 136 miles." (Courtesy Vermilion County Museum.)

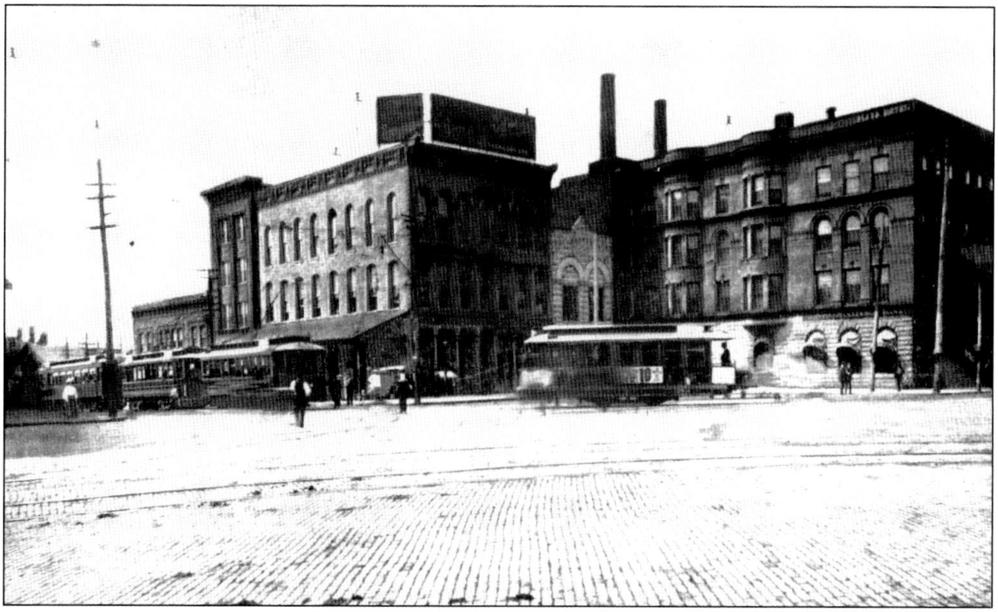

The four corners of the intersection of Main and Vermilion Streets composed the square in Danville. The Plaza Hotel and the interurban station occupied the southwest corner of the square. Original sections of the hotel dated to 1859, and the structure was added onto over the years until it reached 60,000 square feet and contained 155 rooms. The hotel was razed in 1970. The interurban station served passengers until 1923. (Courtesy Vermilion County Museum.)

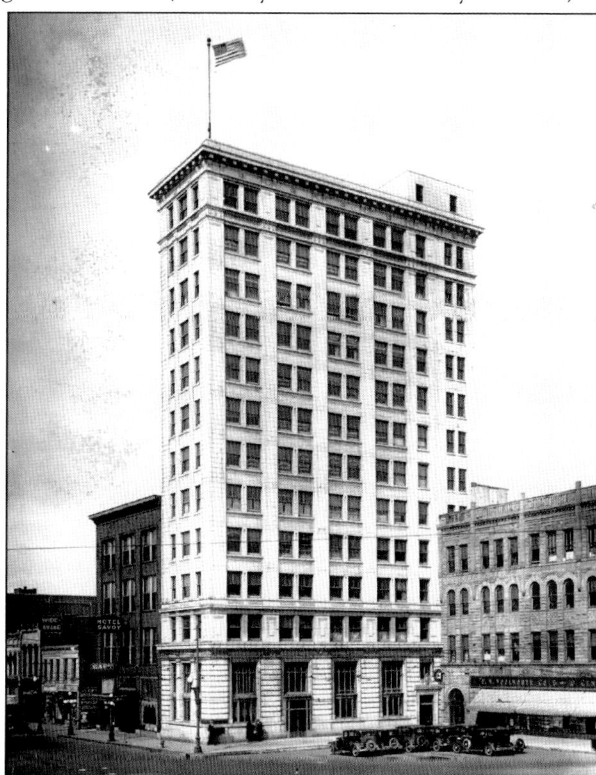

The First National Bank and the F. W. Woolworth buildings occupied the northwest corner of the square in Danville. The 12-story bank building was completed in 1918 at a cost of $350,000 and still stands. The structure has been extensively renovated and serves as an annex to the Vermilion County Courthouse. (Courtesy Vermilion County Museum.)

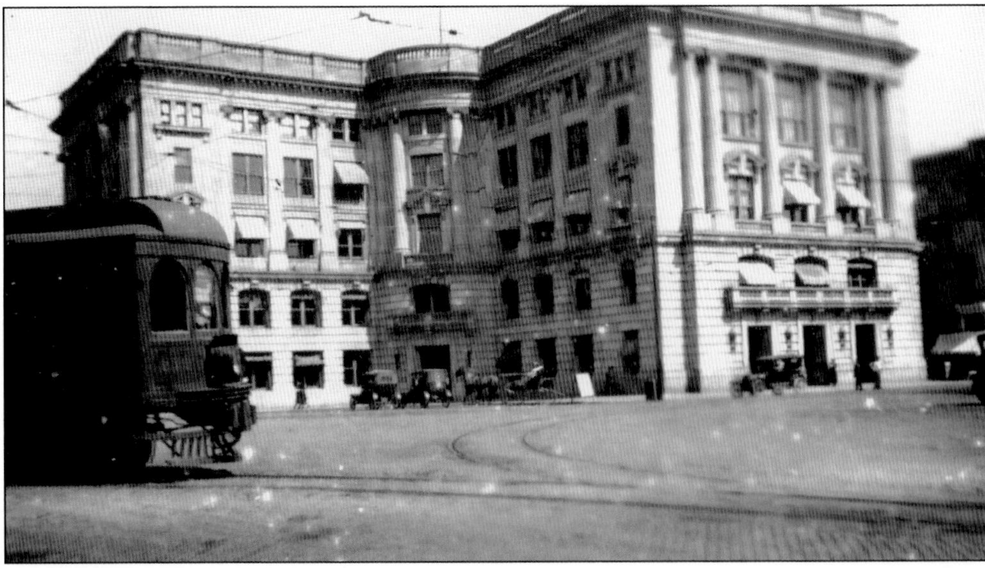

The Vermilion County Courthouse in Danville stands on the northeast corner of the square. The first courthouse was built in 1833 and was destroyed by fire in 1872. The second courthouse was completed in 1876. It was later deemed too small and torn down. The third and current structure, rebuilt on the same footprint as the second courthouse, was completed in 1912. (Courtesy Vermilion County Museum.)

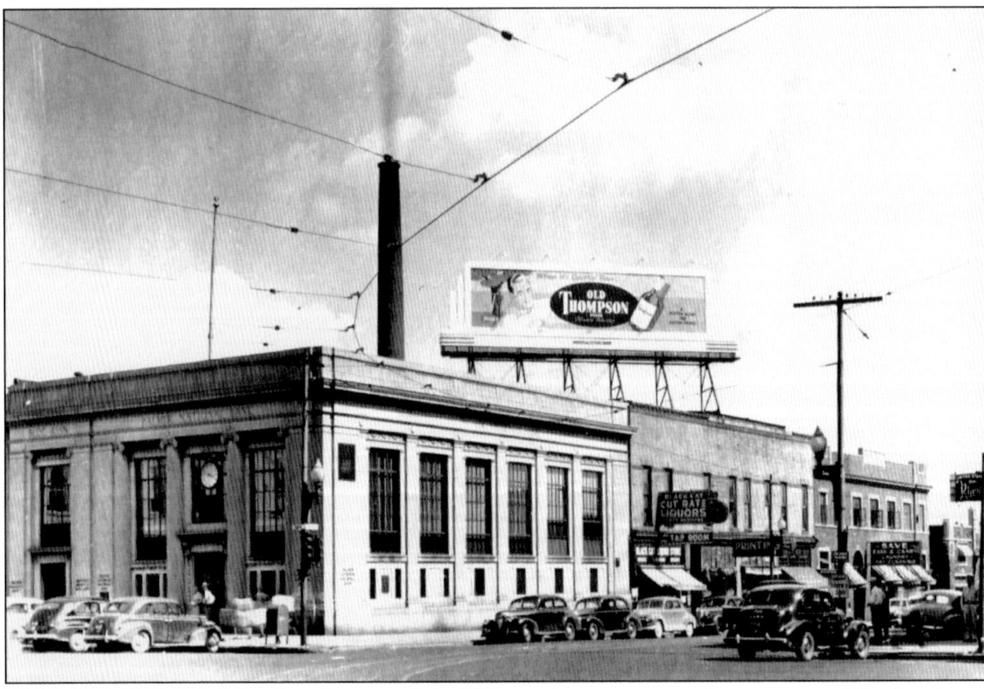

The Palmer National Bank was located on the southeast corner of the square. The bank was operated by the Palmer family, who had been in the financial business in Danville since 1872. The building was constructed in 1911 and was remodeled a number of times over the years until it was eventually torn down in 1976. The site is now a part of the Towne Center parking lot. (Courtesy Vermilion County Museum.)

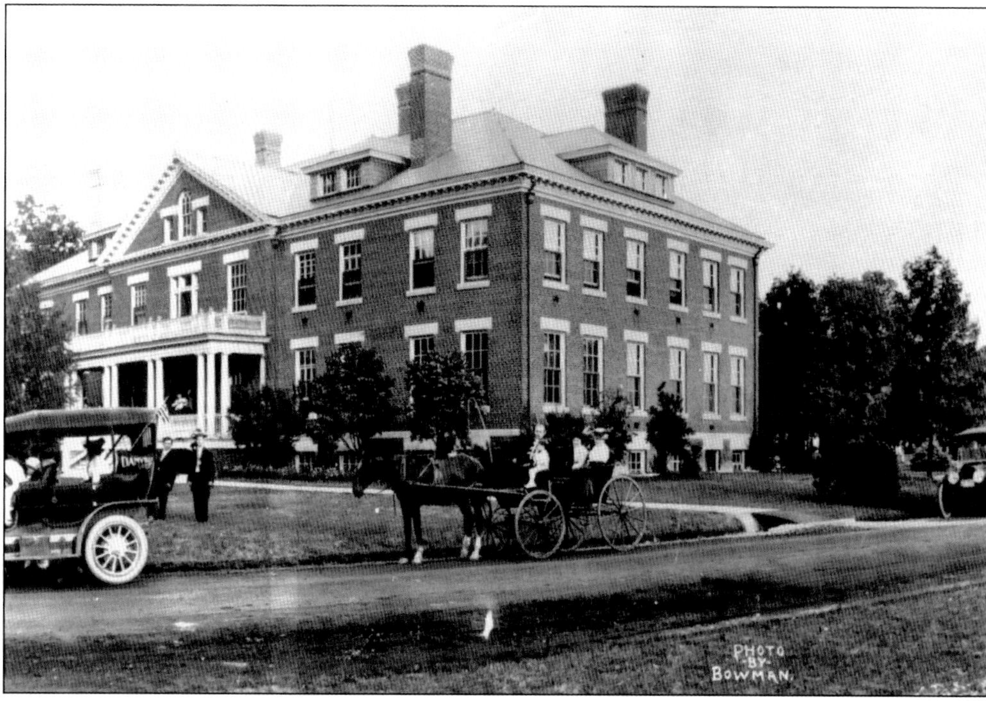

The Danville Soldiers' Home was opened in October 1898 as a residential facility for disabled soldiers and sailors. The home was located on Main Street, about two miles east of downtown Danville. A total of 31 "members" were living at the home at the end of 1898, and at the height of the facility's operations over 20 years later, "membership" had increased to about 4,000 residents. The structure was built in 1899 at a cost of $28,277 and has since been demolished. (Courtesy Vermilion County Museum.)

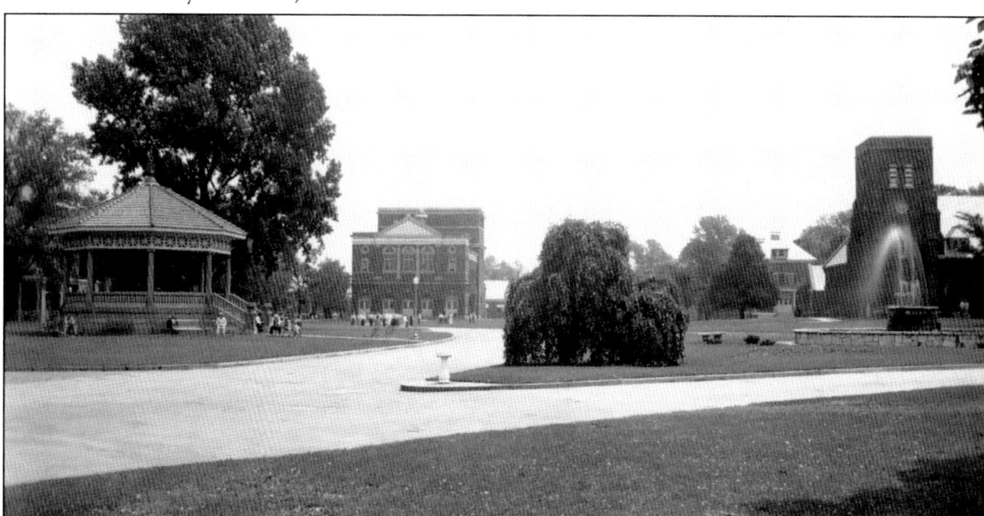

The Danville Soldiers' Home located on the Dixie Highway route east of town was virtually a city in itself. The sprawling complex included beautifully landscaped grounds and amenities such as a band shell, auditorium and theater building, chapel, and fountain. Today the Veteran's Administration Medical Center and the Danville Area Community College occupy the grounds and remaining buildings. (Courtesy Vermilion County Museum.)

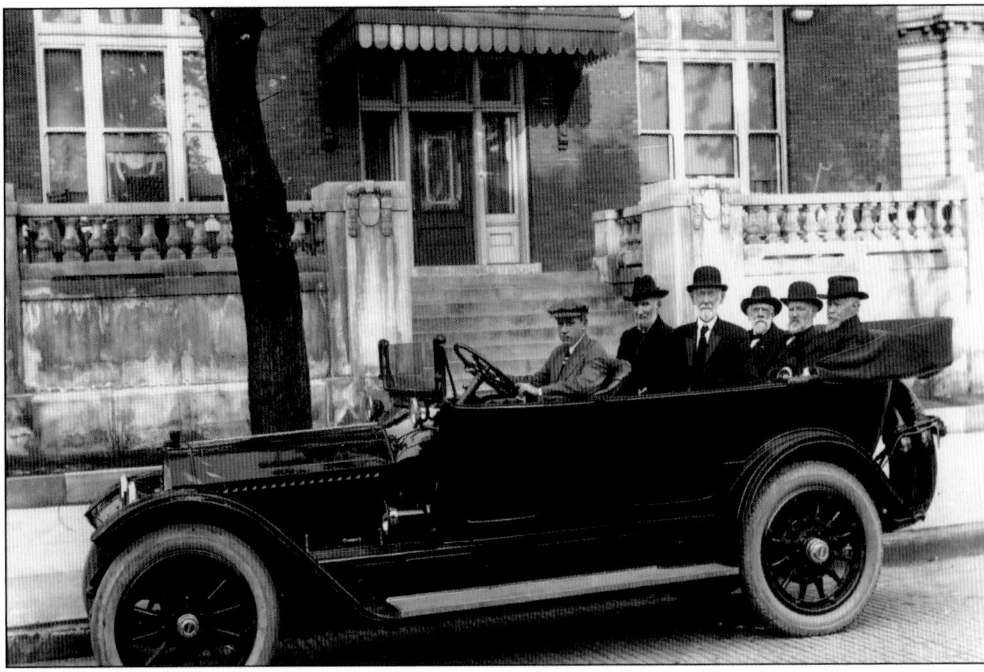

"Uncle Joe" Cannon, second from left, is pictured with other elder Danville elite in front of the Elks Club in the city. Cannon, a resident of Danville, was first elected to Congress in 1873 representing eastern Illinois. Cannon later served 48 years in Congress and as Speaker of the House from 1903 to 1911. Cannon's influence brought much to Vermilion County and Danville, including the soldiers' home, the federal building, and funds for early road development. (Courtesy Vermilion County Museum.)

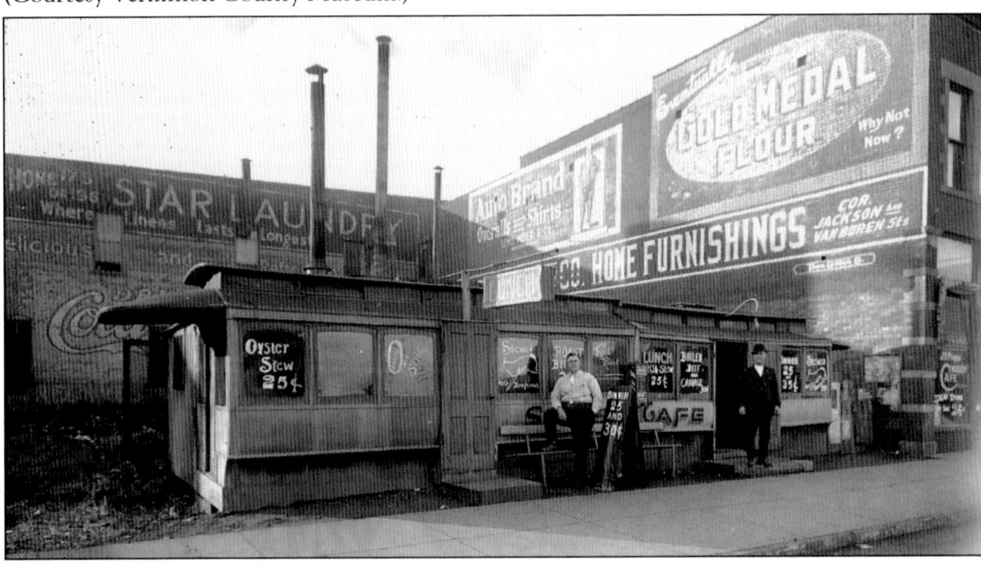

The old-fashioned diner located off East Main Street and Washington Avenue in Danville was a stop for locals and travelers alike for a quick and filling meal. The menu posted on the windows of the retired horse-drawn streetcars that had been cobbled together to form the eatery featured stewed and fried rabbit, Irish stew, boiled beef and cabbage, oyster stew, and roast beef. Lunch and dinner prices ranged between 25¢ and 35¢. (Courtesy Vermilion County Museum.)

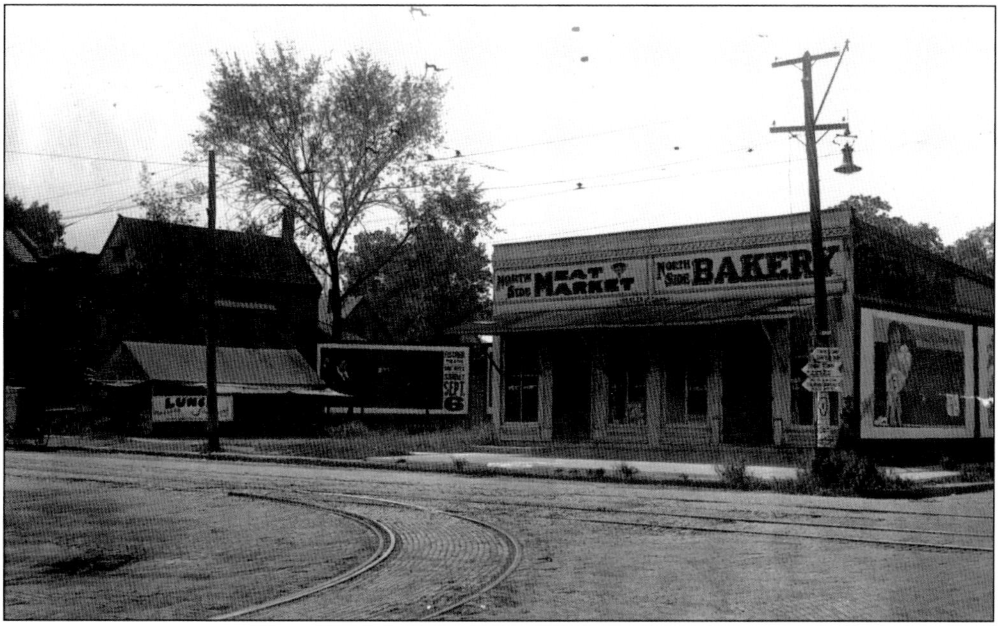

This view of Vermilion Street looking south is at the intersection of Fairchild Street just north of downtown Danville. The North Side Meat Market and Bakery dominate the view, but to the left is a small lunch stand with a sign that proclaims, "Blow horn for curb service." The arrow boards on the light pole in front of the bakery point the direction to downtown, the tourist bureau at the Plaza Hotel, and the tourist camp at Garfield Park. Another pole marking designates Vermilion Street as Illinois Route 1. (Courtesy Vermilion County Museum.)

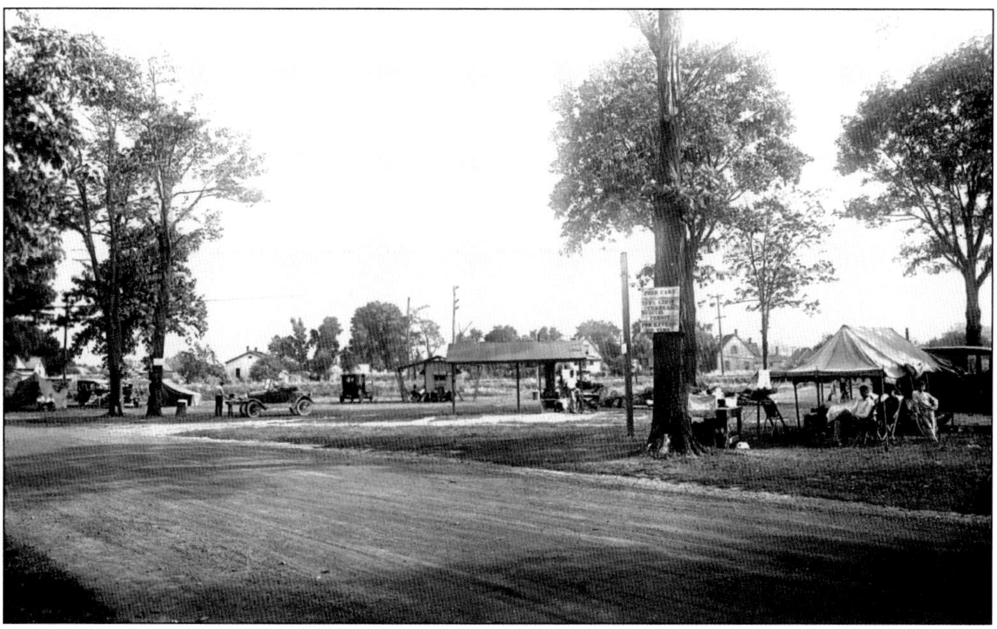

Free camping was available to travelers passing through Danville at Garfield Park, located on North Griffin Street south of Fairchild Street. The simple rules on the sign state that the time limit is three days and a special permit is required for an extended stay. (Courtesy Vermilion County Museum.)

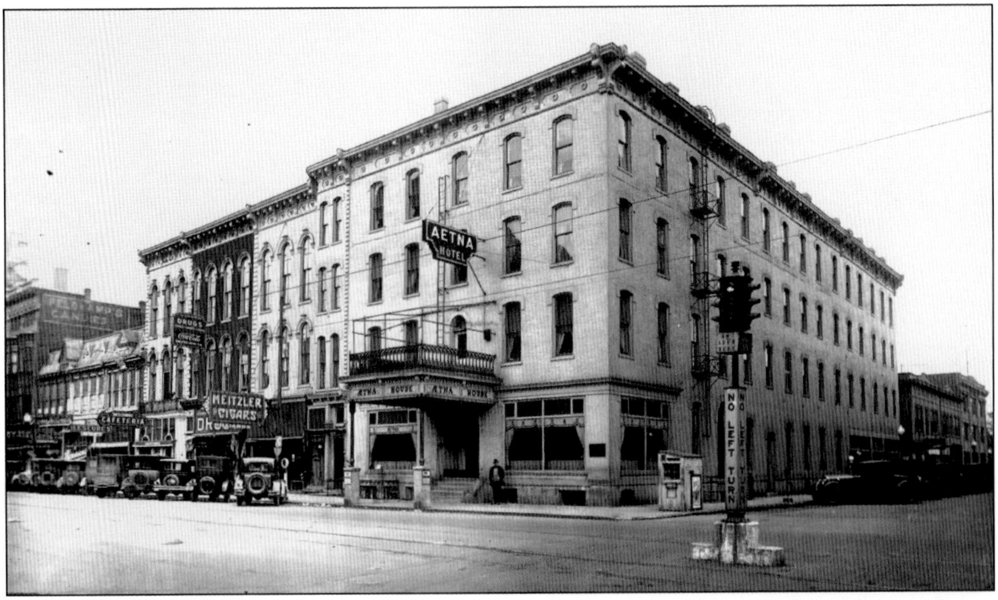

For those not interested in roughing it by camping out at Garfield Park, Danville was home to a number of hotels catering to different budgets. The Aetna Hotel, located at the corner of North and Vermilion Streets, was built in 1863. Just like the hotel, the street corner newsstand and the center-mounted traffic stanchion have faded away with history. (Courtesy Vermilion County Museum.)

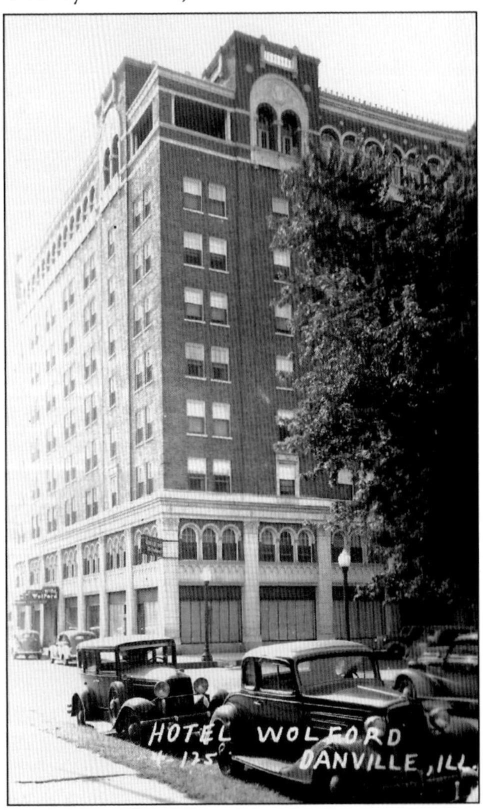

The Hotel Wolford in Danville was completed in 1926 and was once considered to be one of the most elegant in the Midwest. The hotel, located on Harrison Street, just east of the Dixie Highway and Vermilion Street, maintained all the amenities associated with only the finest hotels. Pres. Gerald Ford, Tyrone Power, Danny Kaye, Roy Rogers, Dizzy Dean, and Jack Dempsey were counted as some of the hotel's guests. The hotel closed in 1975 and was converted to senior housing. (Author's collection.)

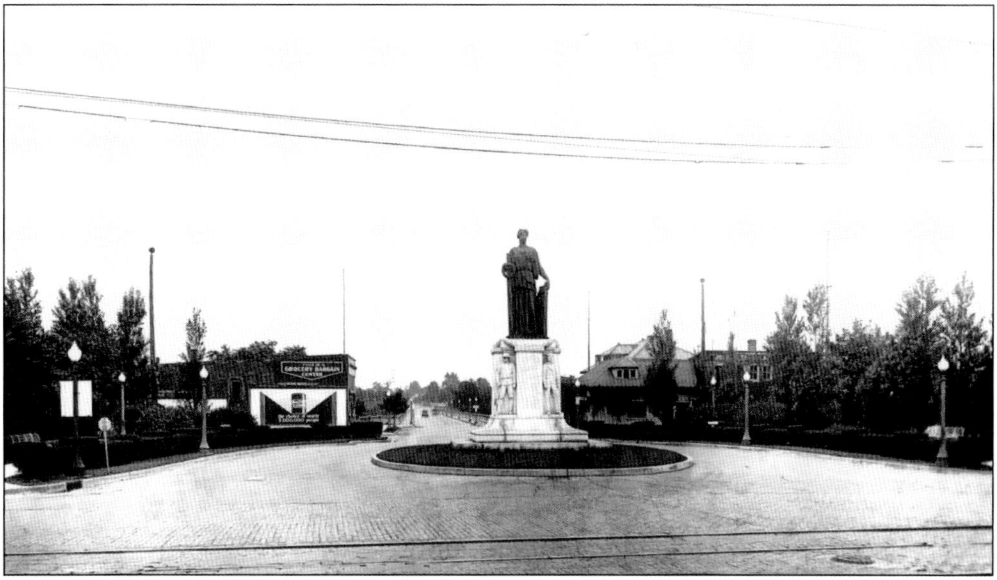

Loredo Taft's *Statue of Victory* was dedicated on November 11, 1922, and was placed at Main and Gilbert Streets at the center of the northern approach to the Victory (now Memorial) Bridge in Danville. By the mid-1930s, downtown Vermilion Street businessmen, concerned by heavy traffic on their street, were successful in having Illinois Route 1 and the Dixie Highway rerouted to Gilbert Street south of Fairfield. In 1955, the statue was relocated to its present site just west of the intersection. (Courtesy Vermilion County Museum.)

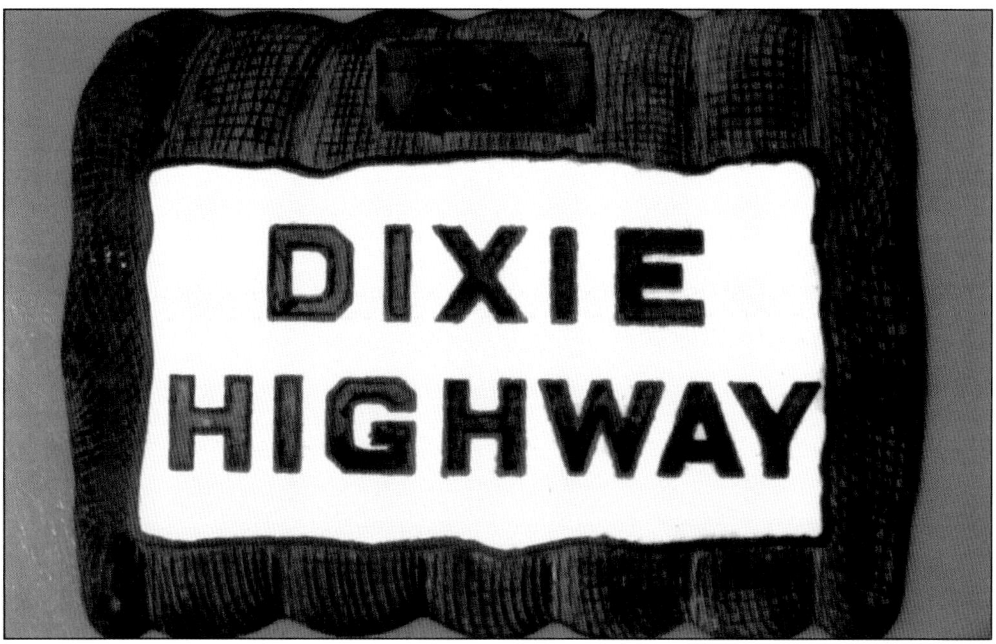

The original logo for the Dixie Highway was a bale of cotton with the words *Dixie Highway* imprinted in the center. The logo was used on road signs and was superimposed on a background of alternating red, white, and red horizontal bands. The cotton bale was a golden brown color, and the print was in black. Within a few years, these colorful signs gave way to the more common red and white DH signs. (Courtesy Carol Triebold.)